Springer-Verlag   6900 Heidelberg 1 · Postfach 1780
Telefon (06221) 49101 · Telex 04-61723
1000 Berlin 33 · Heidelberger Platz 3
Telefon (0311) 822001 · Telex 01-83319

Springer-Verlag   New York, NY 10010 · 175, Fifth Avenue
New York Inc.   Telefon 673-2660

22 Fortschritte der chemischen Forschung
Topics in Current Chemistry

σ and π Electrons in Organic Compounds

Springer-Verlag Berlin Heidelberg GmbH 1971

ISBN 978-3-540-05473-3   ISBN 978-3-540-36655-3 (eBook)
DOI 10.1007/978-3-540-36655-3

Originally published by Springer-Verlag Berlin Heidelberg New York in 1971

Contents

σ and π Electrons in Theoretical Organic Chemistry

W. Kutzelnigg

Institut für Physikalische Chemie der Universität, Karlsruhe

G. Del Re

Cattedra di Chimica Teorica della Facoltà di Scienze di Napoli, Istituto Chimico dell'Università

G. Berthier

Laboratoire de Biochimie Théorique associé au CNRS, Paris

Contents

1

Contents

2

1. Origin and Importance of the $\sigma-\pi$ Separation

1.1. π Electron Theories and the $\sigma-\pi$ Separation

In the development of quantum chemistry, few concepts have proved to be as significant as the distinction between σ and π electrons in organic compounds. This distinction suggested the approximation known as $\sigma-\pi$ *separation* which has made it possible to calculate many important physical and chemical properties of unsaturated compounds within the frame of a 'pure π-electron theory'. This type of theory, which goes back essentially to E. Hückel [1], has the advantage of great conceptual and practical simplicity and has been successful in solving many problems. Nowadays, the advent of computers has made it feasible to treat polyatomic molecules of small and medium size taking into account all the electrons. Nevertheless, scientific economy suggests that, if certain physical or chemical facts can be understood in terms of π electrons only, one should try to do so; therefore, 'π-electron theories' still deserve analysis and applications.

The justification of π-electron theories has been repeatedly questioned during recent years; indeed, it has become almost fashionable to emphasize the shortcomings of the $\sigma-\pi$ separation and the non-validity of the theories based upon it. These are, in fact, approximations and cannot be expected to lead to unconditionally reliable conclusions. However, the numerical results that have provoked the criticisms in question are not a necessary consequence of the $\sigma-\pi$ separation and the related approximations. Therefore, we shall begin by restating and clarifying the basic concepts on which the whole question of the $\sigma-\pi$ separation rests. We shall consider the conditions under which the electrons of a molecule can be classified into σ and π electrons and indicate what should be understood be '$\sigma-\pi$ separation' and what are the limitations of this approximation. We shall show that the most important part of the '$\sigma-\pi$ interaction' is usually taken into account within the $\sigma-\pi$ separation scheme and, finally, discuss whether the $\sigma-\pi$ interaction has a significant effect on the theoretical predictions made for the physical properties of unsaturated molecules (ionization potentials, electronic spectra, charge densities and dipole moments etc.).

To preserve the rigor of certain arguments, a number of quantum-mechanical formulas are useful; they will be introduced, when required, with the necessary explanations of notations.

1.2. Historical Background

The terms 'σ and π electrons' come from quantum mechanics, but the idea that two different types of bonds between, say, carbon atoms should be distinguished occurred in organic chemistry long before the advent of quantum chemistry. A detailed historical review of the entire question is outside the scope of this article (see *e.g.* [2]), and we shall remind the reader only of the milestones in the theories of unsaturated, conjugated, and aromatic compounds.

The most basic notion of organic chemistry is probably the *quadrivalency of carbon*, which was very clearly formulated by Kékulé in 1858 [3]. Olefinic compounds like ethylene suggested that the carbon atom could exhibit the valence three, but these molecules were finally formulated with a double bond, according to Erlenmeyer's proposition [4]. *Kékulé's benzene formula* [5] completed this classic period of valence theory. About 1875, Le Bel [6] and Van t'Hoff [7] introduced the theory of steric valency, where the double bonds between carbon atoms were looked at from a new point of view: Van t'Hoff proposed his famous model, where the tetrahedra of doubly-bonded carbon atoms were supposed to have an edge in common and those of triply-bonded carbon atoms a face in common. This picture was quite satisfactory for isolated double bonds, but the peculiar properties of conjugated and aromatic systems could be understood only by imagining that different double bonds in a molecule can interact in a way not possible for single bonds.

Around 1900, two new theories were developed: Thiele [8] suggested that in a double bond the valencies of the atoms could not be incompletely used and that the residual valencies could interact with each other, as shown in formulas *(1) (2) (3)*

1   *2*   *3*

Nef [9] presented arguments against the universal quadrivalency of carbon and suggested that there are two forms of ethylene in thermal equilibrium, ordinary ethylene *(4)* and an active form *(5)* in which two valencies are unused or 'free'.

$$H\diagdown_{H}C=C\diagup^{H}_{H} \rightleftharpoons H\diagdown_{H}C\cdot\cdot C\diagup^{H}_{H}$$

4   5

These two theories can be regarded as the first realizations that a conjugated system is to be described in terms of two different kinds of bonds, those of the first kind being localized between two neighbouring atoms, those of the second kind extending over several atoms, and that the latter cannot always be represented by a single structural formula.

It is not worthwhile recalling in detail all the facts which were used to support and improve the classical theories of organic chemistry till the development of the quantum theory of the chemical bond; but it is useful to outline the reasons why experimentalists have come to speak of *single* and *double bonds*. The story goes back to the discovery first of the peculiar properties of unsaturated compounds and later of conjugated compounds. Having accepted the quadrivalency of carbon, the chemists found that, whenever a hydrocarbon contained a carbon atom forming several bonds with a single partner, the properties of the molecule were radically different from those of hydrocarbons with four partners per carbon; this is why they were led to consider a double bond as a superposition of two non-equivalent bonds, one of them being quite different from the typical C—C bond of a saturated compound. For instance, ethylene, easily adds a number of molecules, whereas ethane and propane are unreactive; at first sight, it looks as if one of the two bonds linking the carbons together is easily broken, while the other remains in place and behaves as an 'ordinary' bond, *i.e.* as the C—C bond of paraffins. In the case of conjugated systems, like butadiene, the situation is even more surprising; not only is there a difference between, so to speak, the 'first' and the 'second' bond of a double bond, but the various 'second' bonds behave as an entity, thus suggesting that they interact strongly with one another, at variance with the 'ordinary' bond.

In short, the notion that a double bond consists of one bond having properties very similar to those of the corresponding bond in a saturated compound, while the other has very peculiar properties, is suggested, so to speak, by experimental evidence: the 'second' bond seems to be responsible for the high reactivity of olefins, the chemical behaviour of conjugated compounds, the aromaticity of benzene and related molecules, the physical properties characteristic of unsaturated and aromatic compounds etc. . . As a matter of fact, the experimental evidence such as we have just recalled is not really so conclusive as it may seem, especially as far as isolated double bonds are concerned. It is possible to interpret the behaviour of a double bond by saying that such a bond is formed by two 'curved' bonds that, because of the 'strain' to which they are subjected, are relatively weak and hence highly reactive. As soon as one of the 'bent' bonds is broken, the remaining bond becomes straight and hence normal. For isolated double bonds, this picture is as good as the other, and there might even be properties which would be better interpreted in this way; for conjugated double bonds, it does not provide a simple interpretation of their interaction. A more consistent picture, involving a clear distinction between the two types of bonds, was given by quantum mechanics, and the starting point was the study of diatomic molecules by Hund [10] and Mulliken [11] within the frame of the molecular orbital method.

In *diatomic molecules* (as well as in other linear molecules), the internuclear axis is a *symmetry axis* of infinite order, and the molecular orbitals can be classified according to the number of nodal planes (all containing the symmetry axis). Physically, this corresponds to an ordering according to the values of the component of the angular momentum along the symmetry axis. For this classification, Hund [10] proposed the notation σ, π, δ etc.; the σ orbitals have angular momentum zero and no nodal plane, the π orbitals have angular momentum one along the nuclear axis and (if chosen real rather than complex) one nodal plane, and so on. . . It can also be stated (see [11]) that molecules with multiply bonded atoms usually have electrons occupying both σ and π binding orbitals, *i.e.* have one σ bond and one or two π bonds. The few exceptions (like molecules B_2 or C_2), which have no σ bond, but two lone pairs and one or two π bonds respectively) are generally unusual compounds from the chemical point of view. It should be added that π bonds without any underlying σ skeleton are occasionally considered in polyatomic molecules: structures of that sort have been suggested for nitrogen tetroxide [12] and thio-thiophten [13]; a formula with a pure π bond between the two nitrogen atoms of N_2O_4 has been shown to be theoretically incompatible with the observed diamagnetism of this compound [14,15,16].

The success of the preceding scheme for diatomic molecules [17,18,19,20,21] led Hund [22] and Mulliken [23] to apply the same theory to *polyatomic molecules*. In the beginning, there seemed to be no direct relation between molecular orbitals (MO's) and the bonds in a chemical formula, because MO's normally extend over the whole molecule and are not restricted to the region between two atoms. The difficulty was overcome by using equivalent localized MO's instead of the delocalized ones [24,25]. The mathematical definition of equivalent MO's was given only in 1949 by Lennard-Jones and his coworkers [26,27], but the concept of localization

6

is older [22,28,29]. It is not always possible to find a linear transformation that localizes all the electrons properly in bonds, inner shells and lone pairs; only when such is the case can a molecule be described in terms of localized bonds. The condition given by Hund [22] for a localized description was that the number of valence electrons of any atom should be equal to the number of atomic valence orbitals involved in the bonding and to the number of neighbours to which the atom was bound. We shall come back to the localization problem in Section 3.3.

The terminology for the case of diatomic molecules was generalized by considering a polyatomic molecule as a collection of localized bonds. For instance, in ethylene one can speak of a σ bond and a π bond for the double bond between the two carbon atoms. The treatment of benzene presented by Hückel [1] in 1931 followed the same lines: There are $4 \times 6 + 6 = 30$ valence electrons, 24 of which are assumed to participate in six localized CC and CH bonds forming the C_6H_6 frame of the molecule; the remaining six electrons were assigned to MO's constructed from $2p\pi$ atomic orbitals (AO's) whose nodal plane coincides with the molecular plane, and treated independently of the σ electrons. This intuitive introduction of the $\sigma-\pi$ separation is not free from criticism. However, by limiting his treatment to the π electrons, Hückel was able to explain the peculiar properties of benzene and other conjugated and aromatic molecules. The notation 'σ and π orbitals' came from the theory of linear molecules, where such words have a definite meaning. In order to carry them over to non-linear molecules, one has to assume that the bond orbitals are (that is to say, can be) localized between two atoms, because then these orbitals can be classified with respect to linear 'pseudo-symmetry'. This holds both for saturated and unsaturated non-conjugated molecules. On the other hand, in conjugated molecules, the π bonds are delocalized and, strictly speaking, can no longer be classified as π orbitals unless the molecule is completely linear. Nevertheless, unsaturated systems are usually *planar*, so that the orbitals can be classified according to their symmetry with respect to the molecular plane (or, if the molecule is not wholly unsaturated, according to the plane of the unsaturated system). In this way, a redefinition of σ and π orbitals is possible (see section 2.1), but it does not have exactly the same physical meaning as in linear molecules. As a matter of fact, it became customary to speak of σ and π electrons in unsaturated molecules only after 1940 [30]: Hückel [1] used the terms '*Elektronen erster Art*' and '*Elektronen zweiter Art*'; Schmidt [31], another pioneer of the π electron theory, called them A and B electrons.

In many cases, a local planarity in a part of the molecule is sufficient to preserve the concept of σ and π orbitals, provided that the π orbitals can be restricted to the planar part. For instance, the classification of free radicals in σ and π radicals according to the nature of the unpaired electron [32,33] implies only that the molecule has a local symmetry plane with respect to which the singly occupied MO is symmetric or anti-symmetric. As we shall see, 'quasi π orbitals' having many properties in common with genuine π orbitals, except for the 'nodal plane' can be defined even for non-planar systems, e.g. for the benzene molecule during out-of-plane vibrations, or certain reaction intermediates (non-classical carbonium ions).

1.3. References

[1] Hückel, E.: Z. Physik *60*, 423 (1930); *70*, 204 (1931); *72*, 310 (1931); *76*, 628 (1932); *83*, 623 (1933); Z. Elektrochem *.43*, 752 (1937).

[2] Müller, E.: Neuere Anschauungen der organischen Chemie, Berlin–Göttingen–Heidelberg: Springer 1957.

[3] Kékulé, A.: Liebigs Ann. Chem. *106*, 129 (1858).

[4] Erlenmeyer sen. E.: Z. Chem. u. Pharm. 27 (1862).

[5] Kékulé, A.: Bull. Soc. Chim. *3*, 98 (1865); Liebigs Ann. Chem. *137*, 129 (1866).

[6] Le Bel, A.: Bull. Soc. Chim. *22*, 337 (1874).

[7] Van t'Hoff, J.: Bull. Soc. Chim. *23*, (1875); Ber. *10*, 1620 (1877).

[8] Thiele, J.: Liebigs Ann. Chem. *306*, 87 (1899); *308*, 333 (1899).

[9] Nef, J. U.: Liebigs Ann. Chem. *270*, 267 (1892); J. Am. Chem. Soc. *26*, 1337 (1904); *30*, 645 (1908).

[10] Hund, F.: Z. Physik *40*, 742 (1927); *51*, 759 (1928).

[11] Mulliken, R. S.: Phys. Rev. *32*, 186, 761 (1928); *33*, 730 (1929); Z. Elektroch. *36*, 603 (1930).

[12] Coulson, C. A., Duchesne, J.: Bull. Acad. Roy Belg. Cl. Sci. *43*, 522 (1957).

[13] Giacometti, G., Rigatti, G.: J. Chem. Phys. *30*, 1633 (1959).

[14] Green, M., Linnett, J. W.: Trans. Faraday Soc. *57*, 1 (1961).

[15] Brown, R. D., Harcourt, R. D.: Proc. Chem. Soc. page 216 (1961).

[16] Le Goff, R., Serre, J.: Theoret. Chim. Acta *1*, 66 (1962).

[17] Herzberg, G.: Z. Physik *57*, 601 (1929).

[18] Lennard-Jones, J. E.: Trans. Faraday Soc. *25*, 668 (1929).

[19] Wigner, E., Witmer, E. E.: Z. Physik *51*, 859 (1928).

[20] Heitler, W., Herzberg, G.: Z. Physik *53*, 52 (1929).

[21] Slater, J. C.: Phys. Rev. *35*, 509 (1930).

[22] Hund, F.: Z. Physik *73*, 1, 565 (1932); *74*, 1 (1932).

[23] Mulliken, R. S.: Phys. Rev. *40*, 55 (1932); *41*, 49, 751 (1932); *43*, 279 (1933).

[24] Coulson, C. A.: J. Chim. Phys. *46*, 198 (1949).

[25] Lennard-Jones, J. E.: Proc. Roy. Soc. *A 198*, 1, 14 (1949).

[26] Lennard-Jones, J. A., Pople, J. A.: Proc. Roy. Soc. *A 202*, 166 (1950).

[27] Hall, G. G.: Proc. Roy. Soc. *A 202*, 336 (1950).

[28] Pauling, L.: J. Amer. Chem. Soc. *53*, 1367, 3225 (1931); *54*, 988, 3570 (1932).

[29] Slater, J. C.: Phys. Rev. *37*, 481 (1931).

[30] Mulliken, R. S., Rieke, C. A., Brown, W. G.: J. Am. Chem. Soc. *63*, 41 (1941).

[31] Schmidt, O.: Z. Elektrochem. *40*, 211 (1934); *42*, 175 (1936); *43*, 238 (1937); Z. Phys. Chem. *39*, 78 (1938); *42*, 83, 98 (1939); *44*, 191 (1939); *47*, 1 (1940); Naturwissenschaften *26*, 444 (1938); *29*, 146 (1941).

[32] Berthier, G., Lemaire, H., Rassat, A., Veillard, A.: Theoret. Chim. Acta *3*, 213 (1965).

[33] Symons, M. C. R.: J. Chem. Soc. 2276 (1965).

2. Differences between σ and π Electrons

2.1 General Quantum-Mechanical Formulation

The quantum-mechanical equations for a many-particle system (for more details, see e.g. [1,2]) are deduced from the equations of *classical mechanics* by replacing the physical quantities appearing in them (position, momentum etc. . .) by appropriate operators; the latter operate on certain functions, called *wave functions*, which describe the possible states of the system. The values of physical observables are '*the expectation values*' of the corresponding operators. For instance, the expression

$$<\Omega> \equiv <\Psi|\Omega|\Psi> \equiv (\Psi, \Omega\Psi) \equiv \int \Psi^*(\Omega\Psi) \, d\tau \qquad (2.1)$$

is the expectation value of the operator Ω, the three formulas on the left being just *different symbolic ways of writing* the integral on the right. This expression means that one has to apply the prescription corresponding to the operator Ω (multiplication by a coordinate, derivation etc. . .) to the wave function Ψ, multiply by Ψ^*, the complex conjugate of Ψ, and integrate over the whole space of definition for Ψ.

If, as we shall always assume in the following, the variables on which the wave function depends are the $3n$ position coordinates x_1, y_1, z_1, . . ., x_n, y_n, z_n of the n particles of the given system, the volume element for integrating is $d\tau = dx_1 \cdot dy_1 \cdot dz_1 \cdot \ldots \cdot dx_n \cdot dy_n \cdot dz_n$ (In principle, one should also consider the so-called 'spin coordinates'; they will be explicitly introduced as the need arises).

The fundamental *Hamiltonian operator H*, whose expectation values give the energies of the possible states of an atom or a molecule, is the sum of the operator T corresponding to the total kinetic energy and the operator V corresponding to the mutual potential energy of electrons and nuclei and, if an external field is present, the potential energy of the system in that field. Because of their larger mass, the nuclei move much more slowly than the electrons; therefore, the Born-Oppenheimer approximation can be introduced, that is to say, the *nuclear coordinates can be treated as fixed parameters*. Then, for a given configuration (usually, the equilibrium one) the nuclei appear in the equation of motion only

9

as the sources of an external electrostatic field acting on the electrons, and their mutual potential energy can be added to the electronic energy as a constant term to compute the total molecular energy. In brief, for an isolated molecule, the Hamiltonian operator giving the electronic and nuclear energies has the form

$$H = T + V$$

$$T = \frac{1}{2} \sum_{\nu=1}^{n} \left(\frac{\partial^2}{\partial x_\nu^2} + \frac{\partial^2}{\partial y_\nu^2} + \frac{\partial^2}{\partial z_\nu^2} \right) \tag{2.2}$$

$$V = - \sum_{K=1}^{N} \sum_{\nu=1}^{n} \frac{Z_K}{r_{K\nu}} + \sum_{\mu=1}^{n} \sum_{\nu=1}^{\mu-1} \frac{1}{r_{\mu\nu}} + \sum_{K=1}^{N} \sum_{L=1}^{K-1} \frac{Z_K Z_L}{r_{KL}}$$

the indices μ and ν refer to electrons,
the indices K and L to nuclei;
$r_{K\nu}$, for instance, is the distance of the νth electron from the Kth nucleus.
N and n are the numbers of nuclei and electrons, respectively,
Z_K denotes the positive charge of the Kth nucleus.

In Eq. (2.2), all the operators are expressed in atomic units, so that the physical constants (mass and charge of electron etc...) are omitted.

The system of atomic units is defined by the rest mass of electron m (unit of mass), the magnitude of the charge on electron e (unit of electron charge), the radius of the first Bohr orbit of hydrogen atom a_0 (unit of length and the modified Planck constant $\hbar = h/2\pi$ (unit of angular momentum). The corresponding unit of energy e^2/a_0 is twice the ionization potential of hydrogen. The most recent values of physical constants [3] give for the unit of length (called also Bohr: B) and for the unit of energy (called also Hartree: H) the following correspondence in the CGS system:

$a_0 = 0.529167 \; 10^{-8}$cm, $e^2/a_0 = 219474$ cm^{-1} (27.2107 eV or 627 kcal.mol^{-1}).

The lowest energy state or '*ground*' *state* of a system is the one for which the expectation value of H reaches its absolute minium. More generally, the allowed energies of a conservative system correspond to wave functions Ψ making the energy expectation value stationary. These functions are then given by the solutions of the '*time-independent Schrödinger equation*'

$$H\Psi = E\Psi \tag{2.3}$$

subject to normalization and other appropriate boundary conditions. Such conditions can be satisfied only if the constant E in Eq. (2.3) takes special values, namely the eigenvalues E_i of H, which give the allowed energies of the system. The corresponding solutions of Eq. (2.3) are the normalized eigenfunctions Ψ_i of H.

The normalization condition of the wave function Ψ is

$$\int \Psi^*\Psi d\tau = 1 \tag{2.4}$$

where the integral is taken over the whole space of definition of Ψ. An interpretation of it can be given by considering the quantity $\Psi^*\Psi\, d\tau$ (where the product $\Psi^*\Psi$ is a function of the position coordinates of all the electrons, as is Ψ itself) as the probability of finding at a given time the first electron with coordinates falling between x_1 and $x_1 + dx_1$, y_1 and $y_1 + dy_1$, z_1 and $z_1 + dz_1$, the second electron with coordinates falling between x_2 and $x_2 + dx_2$, y_2 and $y_2 + dy_2$, z_2 and $z_2 + dz_2$, etc. So, the integral appearing in Eq. (2.4) represents the probability of finding the n electrons anywhere in space and must be equal to unity. The product $\Psi^*\Psi$ is called the '*probability density*' for finding the electrons at the position specified by the values of the coordinates for which $\Psi^*\Psi$ is calculated. As we are interested only in the electrons of a molecule, we take Eq. (2.3) with H given by (2.2) as the equation for the (electronic) states of the molecule.

If a molecule has certain *symmetry properties*, important *predictions* about the solutions of the electronic Schrödinger equation can be made without having to solve the equation itself. Consider the case of a planar molecule, *i.e.* of a molecule whose nuclei lie in a plane. This plane is a symmetry plane for the molecule, and it can be shown that any eigenfunction is either symmetric or antisymmetric with respect to this plane. If one chooses the plane of the nuclei as the (y, z) plane of a Cartesian coordinate system, this means that

$$\Psi(x_1, y_1, z_1, \cdots, x_n, y_n, z_n) = \pm\, \Psi\,(-x_1, y_1, z_1, \cdots, -x_n, y_n, z_n) \tag{2.5}$$

If there is more than one symmetry element (symmetry plane, axis etc.), relations similar to Eq. (2.5) hold for every element, and the wave functions can be classified according to group-theoretical symbols (see e.g. [4]).

2.2 One-Electron Molecules and Orbitals

Let us consider more specifically wave functions depending on the coordinates (and possibly on the spin) of a single electron. Such functions are called '*orbitals*' (or, if spin is explicitly included, 'spin orbitals'). According to their behaviour under a reflection with respect to the nuclear plane of a planar molecule, they are classified as σ or π orbitals; one has

$$\begin{aligned} \varphi\,(x_1,\, y_1,\, z_1) &= \varphi\,(-x_1,\, y_1,\, z_1) &&\text{for } \sigma \text{ orbitals} \\ \varphi\,(x_1,\, y_1,\, z_1) &= -\,\varphi\,(-x_1,\, y_1,\, z_1) &&\text{for } \pi \text{ orbitals} \end{aligned} \tag{2.6}$$

The eigenfunctions of the Schrödinger equation for planar one-electron systems, *e.g.* those of the H_3^{++} ion in a non-linear configuration, must be either σ or π orbitals; the former are symmetric about the molecular plane, the latter antisymmetric. The σ orbitals have in general maximum values close to the nuclei, whereas the π orbitals have a nodal surface on the nuclear plane and different signs on the two sides of this plane, as can be seen from Eq. (2.6) by letting $x_1 \to 0$.

Orbitals can also be defined for many-electron systems, and the molecular orbital theory mentioned in the previous section is indeed based on this possibility. In order to assess the significance and limitations of the molecular orbital scheme and the meaning of σ and π orbitals, we have to discuss the definition and the determination of orbitals in a many-electron system at some length.

2.3. Electron Densities and Orbitals in Many-Electron Systems

In order to define orbitals in a many-electron system, two approaches are possible, which we may refer to as 'constructive' and 'analytic'. The first approach is more common: one makes the *ad hoc* postulate that every electron can be associated with one orbital and the total wave function can be constructed from these orbitals. Then, one is led to an 'effective' one-electron Schrödinger equation from one electron in the field of the other electrons. The underlying model is the *'independent particle model'* (IPM). When following the constructive way, one does not know *a priori* whether the model is a good approximation to the actual physical situation; one only knows that it cannot be rigorously correct. The merit of this approach is its relative simplicity from both the mathematical and physical points of view.

In *the analytic approach* one assumes that the state under consideration is described by a sufficiently good wave function and tries to interpret that wave function in terms of orbitals. The first step in this approach is to construct the *electron density*, which is obtained by integrating the probability density $\Psi^*\Psi$ over the coordinates of all particles but the first:

$$\varrho(x_1, y_1, z_1) = n \int \Psi\Psi^* \, dv_2 \ldots dv_\nu \ldots dv_n \, ds \qquad (2.7)$$

Here, dv_ν is the volume element $dx_\nu \, dy_\nu \, dz_\nu$ of the ν-th electron and ds stands for integration over all spin coordinates, so that ϱ is by definition independent of spin. In this way, one obtains the probability density for finding the first electron at the point x_1, y_1, z_1, *i.e.* the electron density at that point.

With some linguistic precautions connected with the wave-particle dualism of quantum mechanics, the electron density ϱ can be interpreted as giving the electronic distribution in the molecule. As shown in Fig. 1 for homonuclear diatomic molecules, this function can easily be visualized in ordinary space and possibly compared with experimental distributions resulting from the analysis of X-rays or electron diffraction measurements.

Therefore, it is tempting to formulate the properties of molecules in terms of the ϱ function rather than to refer to a highly abstract many-electron wave function. Unfortunately, just because ϱ is a quasi-classical quantity, it is impossible to base on it a whole exact or approximate treatment of atoms or molecules; its role remains that of a description of the results obtained through the calculation of the wave function Ψ itself[a].

In the case of a one-electron system described by an orbital φ, the density ϱ is simply

$$\varrho(x_1,y_1,z_1) = \varphi(x_1,y_1,z_1)\,\varphi^*(x_1,y_1,z_1) \tag{2.8}$$

In the case of a many-electron system whose electrons are treated as independent particles having individual wave functions φ_i, the *electron density* takes the form

$$\varrho(x_1,y_1,z_1) = \sum_i \nu_i\,\varphi_i(x_1,y_1,z_1)\,\varphi_i^*(x_1,y_1,z_1) \tag{2.9}$$

where ν_i is the occupation number of the orbital φ_i, which in the present case can be equal to zero, one or two. According to the Pauli principle, an orbital is at most doubly occupied.

The electron density is closely related to a more general function, the so-called 'spin-free one-particle density matrix' [6,7,8]. Whereas the electron density is a function of the three coordinates x_1, y_1, z_1, the density matrix is a function of six coordinates, which are conventionally noted $x_1, y_1, z_1, x_1', y_1', z_1'$. In the case of a one-electron system, the density matrix is given by

$$\gamma(x_1,y_1,z_1;x_1',y_1',z_1') = \varphi(x_1,y_1,z_1)\,\varphi^*(x_1',y_1',z_1') \tag{2.10}$$

---

[a] A complete discussion of this question cannot be given here; we confine ourselves to reminding the reader that serious mathematical difficulties appear in attempts to calculate many-particle density matrices directly. This is known as 'the N-representability problem' [5].

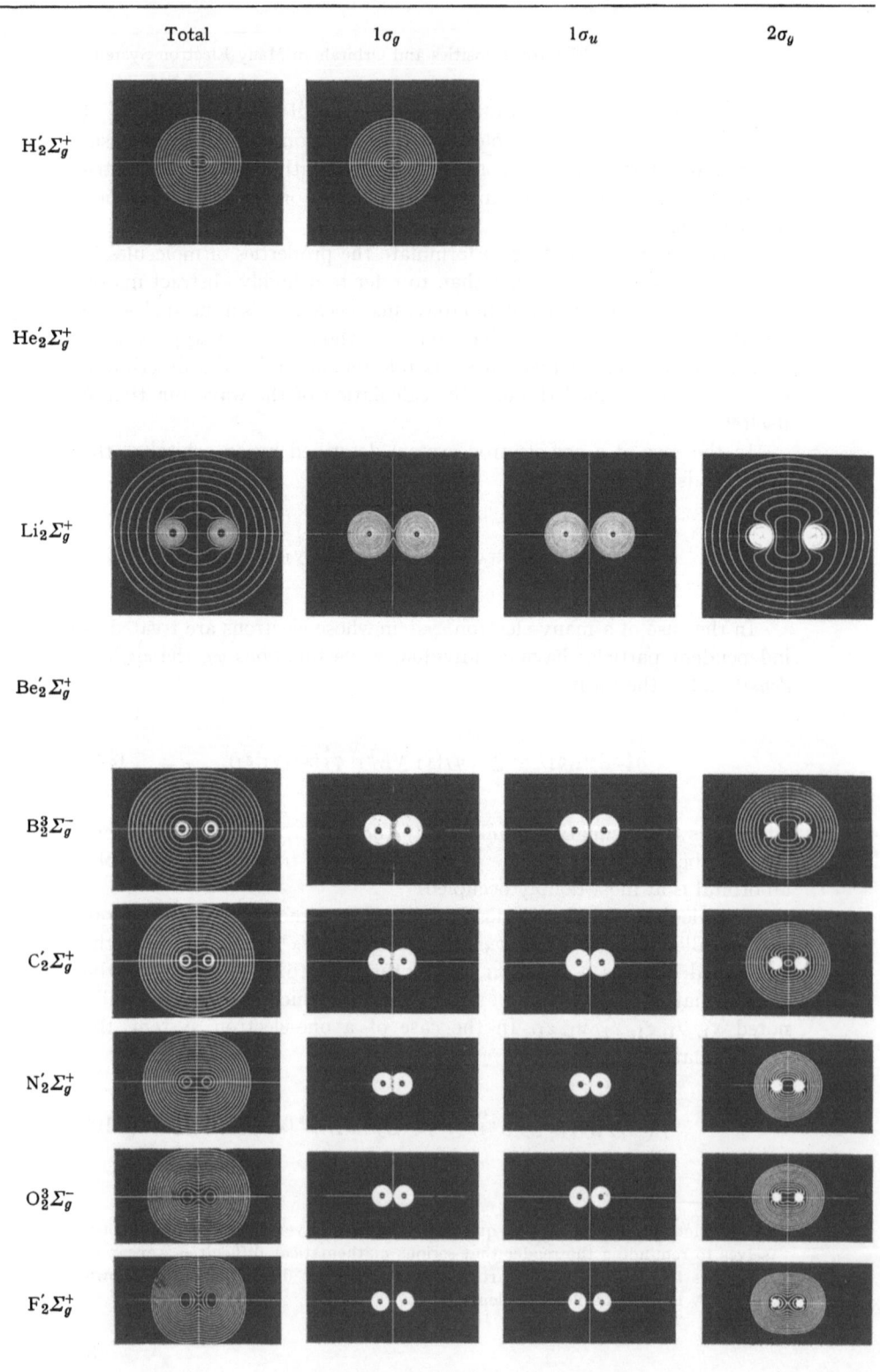

Fig. 1. Total electronic density and orbital densities in homonuclear

$2\sigma_u$ $1\pi_u$ $3\sigma_g$ $1\pi_g$ $3\sigma_u$

For explanation of this figure see next page!

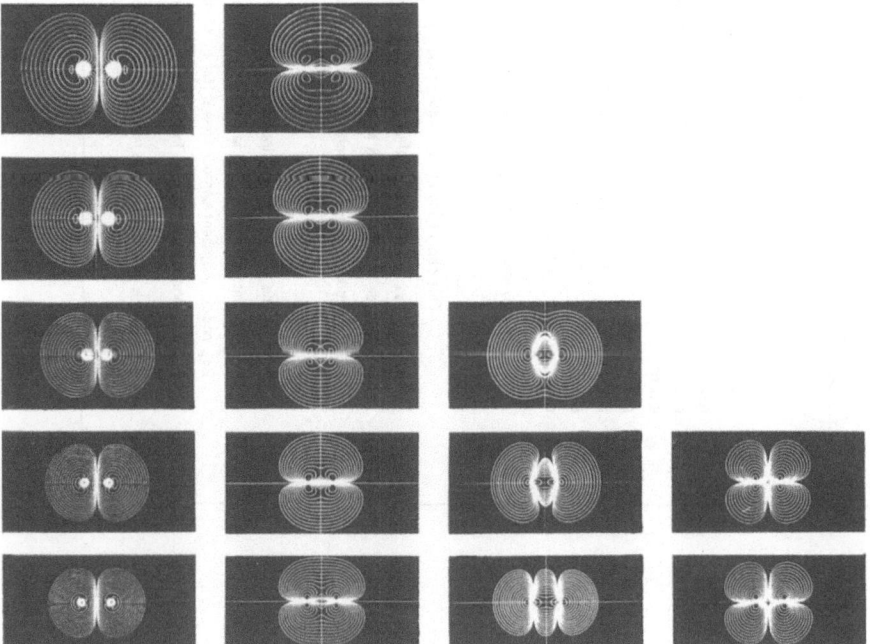

diatomic molecules [after A. C. Wahl: Science *151*, 961 (1966)]

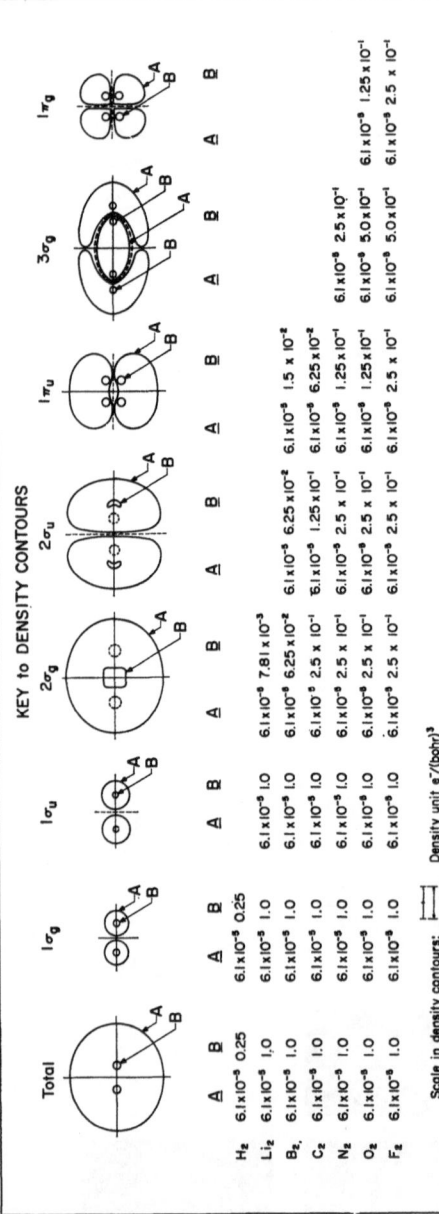

KEY to DENSITY CONTOURS

	Total	$1\sigma_g$	$1\sigma_u$	$2\sigma_g$	$2\sigma_u$	$1\pi_u$	$3\sigma_g$	$1\pi_g$
	A B	A B	A B	A B	A B	A B	A B	A B
H_2	6.1×10^{-5} 0.25	6.1×10^{-5} 0.25						
Li_2	6.1×10^{-5} 1.0	6.1×10^{-5} 1.0	6.1×10^{-5} 1.0	6.1×10^{-5} 7.81×10^{-3}				
B_2	6.1×10^{-5} 1.0	6.1×10^{-5} 1.0	6.1×10^{-5} 1.0	6.1×10^{-5} 6.25×10^{-2}	6.1×10^{-5} 6.25×10^{-2}	6.1×10^{-5} 1.5×10^{-2}		
C_2	6.1×10^{-5} 1.0	6.1×10^{-5} 1.0	6.1×10^{-5} 1.0	6.1×10^{-5} 2.5×10^{-1}	6.1×10^{-5} 1.25×10^{-1}	6.1×10^{-5} 6.25×10^{-2}	6.1×10^{-5} 2.5×10^{-1}	
N_2	6.1×10^{-5} 1.0	6.1×10^{-5} 1.0	6.1×10^{-5} 1.0	6.1×10^{-5} 2.5×10^{-1}	6.1×10^{-5} 2.5×10^{-1}	6.1×10^{-5} 1.25×10^{-1}	6.1×10^{-5} 5.0×10^{-1}	6.1×10^{-5} 1.25×10^{-1}
O_2	6.1×10^{-5} 1.0	6.1×10^{-5} 1.0	6.1×10^{-5} 1.0	6.1×10^{-5} 2.5×10^{-1}	6.1×10^{-5} 2.5×10^{-1}	6.1×10^{-5} 1.25×10^{-1}	6.1×10^{-5} 5.0×10^{-1}	6.1×10^{-5} 2.5×10^{-1}
F_2	6.1×10^{-5} 1.0	6.1×10^{-5} 1.0	6.1×10^{-5} 1.0	6.1×10^{-5} 2.5×10^{-1}	6.1×10^{-5} 2.5×10^{-1}	6.1×10^{-5} 2.5×10^{-1}		

Scale in density contours: $\boxed{}$ 2 bohrs   Density unit $e^-/(bohr)^3$

The SHELL MODEL of MOLECULES

This chart consists of contour diagrams of the electron densities characteristic of the shell model of the molecules H_2, Li_2, B_2, C_2, N_2, O_2, and F_2. Both the total molecular density and the constituent shell densities are displayed at the experimental internuclear distance of each molecule. (He_2, Be_2, and Ne_2 which are members of this homonuclear series are not bound in their ground state and therefore not displayed.)

VALUE of CONTOURS

The above diagrams indicate the general structure of each plot. A labels the lowest contour value plotted and B the highest contour value plotted in each molecule (except for the contours which rise to a value of 1.0 $e^-/(bohr)^3$ inside the $2\sigma_g$ and $2\sigma_u$ node). Adjacent contour lines differ by a factor of 2. Thus all contours plotted are members of the geometric progression $2^{-N} e^-/(bohr)^3$ where N runs from 0 to 14. All plots are in a plane passing through the two nuclei.

This chart was reproduced from "Pictorial Studies of Molecules I: Molecular Orbital Density Comparisons of H_2, Li_2, B_2, C_2, N_2, O_2 and F_2: by Arnold C. Wahl

Layout work by A. H Lung, Graphic Arts, Argonne National Laboratory

Explanation to Fig. 1 on the preceding page

It differs from the density ϱ in the fact that φ and φ^* in the product above are written as functions of different variables. For those values of arguments, where $x_1 = x'_1$, $y_1 = y'_1$, $z_1 = z'_1$, the density matrix γ reduces to ϱ:

$$\gamma(x_1, y_1, z_1; x_1, y_1, z_1) = \varrho(x_1, y_1, z_1) \tag{2.11}$$

If the wave function of a n-electron system is constructed from individual orbitals in the sense of the independent-particle model, γ will have the form.

$$\gamma(1, 1') = \sum_i \gamma_i \varphi_i(1) \varphi_i^*(1) \tag{2.12}$$

completely analogous to Eq. (2.9). In the preceding expression, (1) denotes in short the arguments x_1, y_1, z_1. The general definition of the spin-free one-particle density matrix corresponding to an arbitrary wave function Ψ is the following:

$$\gamma(1, 1') = n \int \Psi(1, 2, 3, \ldots, n) \Psi^*(1', 2, 3, \ldots, n) \, dv_2 dv_3 \ldots dv_n ds \tag{2.13}$$

Note the different notations for a density matrix or an expectation value: the integrand is starred on the right or on the left respectively [8]). It is possible to associate a discrete matrix to the continuous matrix (2.13) by using the fact that any one-electron function, $\varphi_i(1)$, for instance, can be expanded as a linear combination of a given '*complete*' set [b] of one-electron functions χ:

$$\psi_i(1) = \sum_p c_{ip} \chi_p(1) \tag{2.14}$$

If the functions χ form a complete set in one-electron space, *i.e.* in the space of three coordinates, then the products $\chi_p(1) \chi_q^*(1')$ form a complete set in the space of six coordinates. Consequently, γ can be expanded as follows

$$\gamma(1, 1') = \sum_p \sum_q d_{pq} \chi_p(1) \chi_q^*(1') \tag{2.15}$$

---

[b] *Complete sets* are generally infinite; hence, the expansion *(2.14)* contains an infinite number of terms and D is an infinite matrix. However, if the basis set is well chosen, the error made by using a finite number of terms can be made very small — and this is what one has to do in practice.

The coefficients d_{pq} form a matrix in the conventional sense, and γ is completely determined if one indicates the basis functions χ_p and the matrix elements d_{pq}. In practice, γ is mostly given in this way, that is to say, as a matrix *sensu stricto*. In many applications, the atomic orbital basis can be supposed to be orthonormal and the matrix D with elements d_{pq} can be identified with the charge and bond order matrix [7,8].

An expansion of the form (2.15) is possible for any chosen (complete) basis set. It can be stated that there always exists a set of orthonormal functions u_i, *i.e.* a set of functions satisfying the conditions

$$\int u_j^*(1)\, u_i(1)\, dv_1 = \delta_{ji} \tag{2.16}$$

in terms of which a given one-particle density matrix γ is written as

$$\gamma(1,1') = \sum_i \nu_i\, u_i(1)\, u_1^*(1) \tag{2.17}$$

In other words, *for every γ there is a set of orbitals u_i for which the density matrix D is diagonal, i.e.* contains no off-diagonal elements different from zero. These particular orbitals u_i are called the '*natural orbitals*' (NO's) for the state described by the wave function Ψ [7]. In the most general case, the number of functions u_i is not finite, but it can be proved that the occupation number ν_i of any natural orbital lies between 0 and 2:

$$0 \leqslant \nu_i \leqslant 2 \tag{2.18}$$

Numerical calculations show that in usual molecules the occupation numbers ν_i are not exactly integral, but very close to either 0, 1 or 2, so that one can say, at least in first approximation, that a *natural orbital is doubly, singly or not occupied*. This is the main reason why the independent-particle model, defined by putting for convenience the ν_i's equal to 0, 1 or 2, is often a rather good approximation. The occupation numbers of the NO's in a very simple molecule, H_3^+, are given in Table 1.

Another general theorem [9,10] states that the set of NO's associated with every eigenfunction of the Schrödinger equation has definite symmetry properties. In particular, for a planar molecule the natural orbitals are either σ or π orbitals. Therefore, σ and π orbitals have a physical meaning independent of any model assumption or approximation. By adding the occupation numbers of each species of orbitals, one defines

Table 1. *Occupation numbers of the natural orbitals of the molecule-ion in its equilateral form*

N^0	Symmetry species[1]		n_i
1	$1a_1'$	(σ)	0.9825
2,3	$1e'$	(σ)	0.0147
4	$1a_1''$	(π)	0.0014
5	$2a_1$	(σ)	0.0011
6,7	$2e'$	(σ)	0.0002
8,9	$1e''$	(π)	0.0001
10	$3a_1'$	(σ)	0.0000

[1] σ, π with respect to the molecular plane

occupation numbers n_σ and n_π. The sum of all the occupation numbers must be equal to the total number n of electrons; it follows that

$$n = n_\sigma + n_\pi \qquad (2.19)$$

Therefore, one can regard
n_σ as the number of σ electrons and
n_π as the number of π electrons of the molecule

in a given electronic state. In general, these figures will not be integers nor identical with the number of σ and π electrons corresponding to the chemical formula. A simple illustration is given by the molecule H_3^+, where one could conclude from the chemical formula that there are two σ electrons and no π electron, whereas quantum-mechanical calculations [11] lead to $n_\sigma = 1.9971$, $n_\pi = 0.0029$. For typical organic molecules, no accurate values have yet been calculated; in ethylene, n_π will be close to 2, but not exactly equal to 2.

The statement that there is a certain integral number of π electrons in an unsaturated molecule, is a somewhat rough but convenient way of saying that a certain number of π orbitals are 'strongly' occupied, *i.e.* have occupation numbers close to 2. Unlike orbitals, electrons are indistinguishable; therefore, strictly speaking, one should refer to only σ or π orbitals, but never σ or π electrons.

2.4. The Hartree-Fock Model

Current quantum-mechanical calculations are based on the independent-particle model, where one assumes that the molecular orbitals are
either empty or occupied by at most two electrons. This model cannot
give a completely correct description of a many-electron system mainly
because it treats each of the particles as if it 'saw' the others smeared
out in a charge cloud. However, it accounts surprisingly well for many
properties, especially those connected with the one-electron density.
Consequently, it is worthwhile discussing it in detail.

From now on, we shall explicitly use *spin orbitals*, which are derived
from orbitals by multiplying each of them by one of the two possible
spin functions:

$$\xi_{2m-1}(1) = \varphi_m(1)\,\alpha(1)$$

$$\xi_{2m}(2) \quad = \varphi_m(2)\,\beta(2)$$

(2.20)

All the general considerations made so far hold also for spin orbitals,
except for the fact that two spin orbitals may be associated wich one
given orbital, and hence a spin orbital can be occupied at most be one
electron. All the integrals over spin orbitals involve integration over spin
coordinates, and (with limitations which are outside the scope of the
present discussion) this amounts to multiplying the ordinary integral
over the corresponding orbitals by 1 or 0, according to whether or not
the spins are the same [c].

In the frame of the independent particle model, the total wave function can be written as an antisymmetrized product of spin orbitals:

$$\Phi(1,2,\ldots,n) = \frac{1}{\sqrt{n!}}
\begin{vmatrix}
\xi_1(1) & \xi_1(2) & \cdots & \xi_1(n) \\
\xi_2(1) & \xi_2(2) & \cdots & \xi_2(n) \\
\cdots & \cdots & \cdots & \cdots \\
\xi_n(1) & \xi_n(2) & \cdots & \xi_n(n)
\end{vmatrix}$$

(2.21)

The function Φ is known as a '*Slater determinant*'. If one looks for the
energy minimum for such a function, one finds that the orbitals have to
verify the so-called '*Hartree-Fock equations*' [13,14,15]:

$$F\varphi_i = e_i\varphi_i$$

(2.22)

---

[c] In many cases, the conventional spin formulation of quantum chemistry could
be replaced by a spin-free formulation using the permutation symmetry properties
of a n-electron system (see [12]). However, it is then necessary to have recourse
to the complete theory of permutation groups.

with

$$F = T + V_{nuc} + V_{el} \qquad (2.23)$$

where T is the kinetic energy of one electron,
V_{nuc} its potential energy in the field of the bare nuclei, and

V_{el} its potential energy in the averaged field of the other electrons, *i.e.*
for a closed-shell system (inert gases, usual molecules in the ground state
etc.)

$$V_{el}(1) = \sum_i [2 J_i(1) - K_i(1)] \qquad (2.24)$$

J_i and K_i being the *Coulomb* and *exchange operators* corresponding to
each doubly occupied orbital φ_i. The Hartree-Fock equation is an integro-
differential equation which, at variance with a true one-electron Schrödinger equation, involves an operator F depending on the unknown
functions φ_i through the electronic potential V_{el}. Nevertheless, the
operator F can be interpreted as the 'effective' Hamiltonian operator
for one electron in the given molecule. Mathematically, even self-consist-
ency is achieved as regards the potential V_{el}, there is an infinity of different
functions φ_i verifying Eq. (2.22), but only the orbitals φ_i from which
the effective Hamiltonian F is constructed are occupied in the determinant Φ. The other possible solutions are sometimes called '*virtual
orbitals*'; they can be used in first approximation for describing states
of higher energy (see Sect. 5.3).

Eq. (2.22) is much simpler than the original many-electron Schrödinger equation; yet it cannot be solved in closed form and approximation methods must be used. It is customary to choose a finite set of
one-electron basis functions χ and approximate the Hartree-Fock
orbitals φ by an expression similar to Eq. (2.14). If one looks for the
minimum of the total energy given by a wave function constructed from
orbitals of this form, one gets a homogeneous set of linear equations:

$$\sum_q F_{pq}\, c_{iq} = \sum_q e_i\, S_{pq}\, c_{iq} \qquad (2.25)$$

whose coefficients

$$S_{pq} = \int \chi_p^* \, \chi_q \, d\tau$$
$$F_{pq} = \int \chi_p^* \, (F\, \chi_q) \, d\tau \qquad (2.26)$$

are the overlap integrals of the basis functions χ_p and the matrix elements
of the Hartree-Fock operator F, and eigenvalues the orbital energies e_i.

Atomic orbitals are chosen as basis functions in the so-called *LCAO-MO method* [16,17]. However, other choices are possible, for instance, Gaussian functions, which are particularly popular nowadays (see *e.g.* [1]). In planar systems, it is convenient to use basis functions that are either symmetric or antisymmetric with respect to the nuclear plane, *i.e.* are of σ or π species. If χ_p and χ_q are basis atomic orbitals with different symmetry properties, then the matrix element F_{pq} vanishes so that the matrix F is 'factorized' into one σ and one π block[d]:

$$F = \begin{bmatrix} F_{\sigma\sigma} & 0 \\ \hline 0 & F_{\pi\pi} \end{bmatrix} \tag{2.27}$$

This factorization amounts to the statement that Eq. (2.25) breaks down into two separate linear systems, one for the determination of σ orbitals, and the other for π orbitals. In the Hartree-Fock scheme, σ and π orbitals are thus 'separated' simply because the *self-consistent field equations* (SCF equations 2.25) have as solutions φ_i symmetry-adapted functions (*i.e.* in the case of planar unsaturated molecules symmetric or anti-symmetric functions with respect ot the molecular plane), at least for closed-shell ground states [16,18,20,21].

The effective operator for the σ electrons represented by the matrix $F_{\sigma\sigma}$ includes the potential energy of a σ electron in the field of the π electrons, and *vice versa* the effective operator for the π electrons (matrix $F_{\pi\pi}$) includes the potential energy of a π electron in the field of the σ electrons. Even after separation according to Eq. (2.27), Eq. (2.25) is to be solved by an iteration procedure: one guesses the probability distribution of the σ electrons, constructs the matrix $F_{\sigma\sigma}$, calculates the π orbitals, constructs the matrix $F_{\pi\pi}$, calculates new σ orbitals, and so on until the results become stable. Of course, a SCF scheme has to be applied also within each electron group (σ or π).

After these remarks, it may seem that the $\sigma-\pi$ separation in a many-electron system is no more than a formal factorization of the equations governing the independent-particle-model approximation. Actually, the preceding results are of much practical importance, because they imply

---

[d] In fact, symmetry requirements on molecular orbitals introduce in a variational calculation certain constraints, which raise the total energy [18]. This problem, called the '*symmetry dilemma*', has been studied for some π electron systems [19]. It is not important for the present discussion because for a system of closed-shell type the NO's associated with a total wave function of correct symmetry are automatically symmetry-adapted.

that there is a class of molecular orbitals, say the π orbitals, which can be built up from a special type of atomic orbitals (or other basis functions) and treated without detailed information concerning the other class, the role of the latter being just that of creating an effective field. Of course, the effective field can be calculated only if the generating electron distribution is known; however, one may expect that a sufficiently good approximation to it can be found in a rather simple way (as it is an average field) and the results will not depend very much on the precise form assumed for the potential of the σ electrons. This is why one can go even one step beyond the σ-π separation and consider the π electrons only. If one supposes that the field created by the σ electrons has the same general features in a number of unsaturated molecules, one can attribute certain properties to the π orbitals, that is to say to the π electrons, and thus explain the behaviour of molecules just by reference to the π electrons.

Table 2. *Hierarchie of approximate quantum-mechanical theories of unsaturated molecules*

Rigorous non-adiabatic treatment with relativistic corrections
Rigorous solution of the non-relativistic Schrödinger equation in the Born-Oppenheimer approximation
σ-π separation (neglect of intergroup correlation effects) $\Psi_i = \mathscr{A} \{ \Sigma_i \, (1, n_\sigma) \, \Pi_i \, (n_\sigma + 1, \ldots, n_\sigma + n_\pi) \}$ (*i* refers to different spectroscopic states)
Rigid σ core approximation $\Psi_i = \mathscr{A} \{ \Sigma_0 \, (1, \ldots n_\sigma) \, \Pi_i \, (n_\sigma + 1, \ldots, n_\sigma + n_\pi) \}$
Treatment of π electrons in the effective field of the 'core'. Explicit electron interaction within the π group.
Hückel type theories. No explicit interaction within the π group.

In order to understand the question properly, one has to realize that the way from rigorous quantum mechanics to a theory treating only π electrons includes several steps. One step, outlined in Sect. 2.3, is rigorously possible for planar systems and approximately so for locally planar systems; it consists in defining σ and π orbitals and stating what should be understood by 'σ and π electrons'. Another step is the $\sigma-\pi$ separation in the sense just explained, namely the factorization of the Hartree-Fock matrix and its consequences. The $\sigma-\pi$ separation can be formulated in a more general way (not based on the independent-particle model), and we shall consider such a formulation in the next section. Whereas the distinction into σ and π electrons is rigorous, the $\sigma-\pi$ separation is an approximation which is not always a very good one, as will be shown in Sect. 3.2.

In order to formulate a *theory of π electrons only*, additional approximations are necessary: one of them is the assumption of the '*rigid σ core*', another is the Goeppert-Mayer and Sklar potential, which will be discussed in Sect. 5.1

2.5. References

[1] Kutzelnigg, W.: Angew. Chem. *78*, 789 (1966); Int. Ed. *5*, 823 (1966).

[2] Hanna, M. W.: Quantum Mechanics in Chemistry. New York: Benjamin 1965.

[3] Cohen, E. R., DuMond, J. W. M.: Rev. Mod. Phys. *37*, 590 (1965).

[4] Mulliken, R. S.: J. Chem. Phys. *23*, 1997 (1955).

[5] Coleman, A. J.: Rev. Mod. Phys. *35*, 668 (1968).

[6] Husimi, K.: Proc. Phys. Math. Soc. Japan *22*, 264 (1940).

[7] Löwdin, P. O.: Phys. Rev. *97*, 1474, 1490 (1955).

[8] Mc Weeny, R.: Proc. Roy. Soc. (London) *A 253*, 242 (1959); Rev. Mod. Phys. *32*, 335 (1960).

[9] Kutzelnigg, W.: Z. Naturforsch. *18*a, 1058 (1963).

[10] McWeeny R., Kutzelnigg, W.: Int. J. Quant. Chem. *2*, 187 (1968).

[11] Kutzelnigg, W., Ahlrichs, R., Labib-Iskander, I., Bingel, W.: Chem. Phys. Letters *1*, 447 (1967).

[12] Matsen, F. A.: J. Am. Chem. Soc. *92*, 3525 (1970).

[13] Hartree, D. R.: Proc. Cambr. Phil. Soc. *24*, 328 (1928).

[14] Fock, V.: Z. Physik *61*, 126 (1930).

[15] Slater, J. C.: Phys. Rev. *32*, 339 (1928); *35*, 1210 (1930).

[16] Roothaan, C. C. J.: Rev. Mod. Phys. *23*, 69 (1951); *32*, 179 (1960).

[17] Del Re, G.: Int. J. Quant. Chem. *1*, 293 (1967).

[18] Löwdin, P. O.: Rev. Mod. Phys. *35*, 496 (1963).

[19] Paldus, J., Čižek, J.: J. Chem. Phys. *47*, 3976 (1967); *52*, 2919 (1970); *53*, 821 (1970).

[20] Delbrück, M.: Proc. Roy. Soc. *A 129*, 686 (1930).

[21] Adams, W. H.: Phys. Rev. *127*, 1650 (1962).

3. The $\sigma-\pi$ Separation and the Role of Electron Correlation

3.1. $\sigma-\pi$ Separation and Group Function Formalism

In the preceding section, the $\sigma-\pi$ separation occurs as a direct result of the independent-particle model. The derivation is straightforward, but not entirely satisfactory, because the independent-particle model is a simplification of the actual quantum-mechanical situation. In fact, the $\sigma-\pi$ separation can be introduced in the frame of more general treatments, e.g. the 'separated-group function' formalism [1,2,3].

A quantum-mechanical n-electron system is said to consist of M separated groups (A, B, etc.) containing n_A, n_B, etc. electrons, respectively, if it can be described by a wave function of the form

$$\Psi(1,2,\ldots,n) = \mathscr{A}\{\Psi_A(1,2,\ldots,n_A)\,\Psi_B(n_{A+1},n_{A+2},\ldots,n_A+n_B)\ldots\}$$

$$(3.1)$$

where $\mathscr{A}$ is an *antisymmetrization operator* whose purpose is to make Ψ independent of the order of the electrons, and where Ψ_A, Ψ_B, etc. are wave functions of the groups A, B etc... It is customary to impose the so-called 'strong-orthogonality condition'

$$\int \Psi_A^*(1,2,\ldots)\,\Psi_B(1,2,\ldots)\,dv_1 = 0 \qquad (3.2)$$

This condition implies that the natural orbitals of the different groups are mutually orthogonal and exclusive, i.e. no two groups have any (occupied) NO in common, and the NO's of different groups are orthogonal to each other[a].

Physical systems cannot be rigorously described by a separated-group wave function, but that description may often be a rather good approximation. If this is the case, an important simplification of the quantum-

---

[a] This condition is much stronger than conventional orthogonality, because the integral (3.2) should already vanish when taken over the coordinates of one particle, whereas conventional orthogonality means it vanishes after integration over the coordinates of all particles.

mechanical equations is achieved. First, one finds that the *total energy* can be written in the form

$$E = \sum_R E_R + \sum_{R<S} E_{RS} \tag{3.3}$$

where E_R is the energy associated with the Rth group and E_{RS} the interaction energy between the Rth and S^{th} groups. It is important that in such a theory the *'interaction energies'* E_{RS} are expressed in terms of the electron densities (or rather the one-particle density matrices) of the respective groups. The interaction between the groups is essentially electrostatic, but the electron interaction may take a very complicated form within each group. The total electron density is just the sum of the group densities.

'Effective' Schrödinger equations can be derived for the different groups. The presence of group B is reflected in the effective Schrödinger equation for group A only through an effective field (a one-particle potential) due to the charge distribution of group B.

A particularly simple case of a wave function describing separated groups is the single *Slater determinant* of the independent particle model (see Sect. 2.4). There, each group consists of a single electron described by a single orbital. The effective Schrödinger equation for one particle is, of course, the Hartree-Fock equation discussed in Sect. 2.4. Within this model the true *Coulombic interaction* of the electrons is replaced by the interactions of the charge distributions of the orbitals corresponding to the different electrons. That part of the interaction which is ignored in this model, and which comes from the fact that the electrons are not simply 'smeared-out' charge distributions, is called *'electron correlation'*. By definition, electron correlation is completely neglected in the independent-particle model, whereas a wave function of the type (3.1) can account for electron correlation within each group, but neglects electron correlation between electrons in different groups.

If two groups of electrons are well separated in space (*i.e.* far from each other), then it is a very good approximation to identify the real interaction by that of the corresponding charge distributions. Therefore, one can assume that neglect of 'intergroup correlation' is justified if the two groups describe, for instance, two localized bonds far from each other. One can take electron correlation into account by performing a *configuration interaction (CI) calculation*. This is done as follows: one chooses a (more or less arbitrary) set of m spin orbitals and constructs all the $\binom{m}{n}$ n-electron Slater determinants Φ_i that can be obtained from the given spin orbitals. Then, a trial function is written as a linear combination of these determinants

$$\Psi = \sum_i a_i \Phi_i \tag{3.4}$$

and the coefficients a_i are taken as variational parameters. In practice, this procedure is limited by the fact that the number of possible determinants is very large for systems with many electrons. However, in the separated-group method, it is only necessary to perform a configuration interaction for each group independently, which is much easier, provided that the groups are sufficiently small.

The separability condition (3.2) is automatically fulfilled if the orbitals used for the different groups belong to *different symmetry species*. For a planar molecule, the natural way of constructing two separate groups consists in making one group from σ orbitals and another group from π orbitals; the corresponding wave function is

$$\Psi = \mathscr{A} \{\Sigma(1,2,\ldots,n_\sigma)\, \Pi(n_\sigma+1,n_{\sigma+2},\ldots,n_\sigma+n_\pi)\} \qquad (3.5)$$

According to Lykos and Parr [3], an unsaturated molecule can be described in this manner by taking for Σ a (in principle complete) linear combination of Slater determinants built from σ orbitals only and for π a similar combination built from π orbitals only. Such a description is more general than the independent-particle model, as it includes the latter as a special case (namely where both Σ and Π are single Slater determinants).

Of course, the correct wave function cannot be written exactly in the simple form (3.5); whether that expression is a good approximation can hardly be decided *a priori*. Such a wave function is obviously able to account in part for the so-called *'horizontal' electron correlation*, namely for that part of the correlation of the π group is which accounted for by CI with π orbitals only, and *vice versa* for that part of the correlation in the σ group which is accounted for by CI with σ orbitals. Cases where residual contributions to correlation ('vertical' correlation) may play a role will be discussed in Sect. 3.2 [b], in connection with the problem of spin densities.

However, such cases are exceptional, because they are found in highly sophisticated experimental techniques. This is fortunate since practically all the calculations carried out so far, including simultaneous treatments of σ and π electrons, have been based on the σ—π separation, most of them on the much more restrictive independent-particle model.

If we assume the σ—π separation, the total occupation numbers n_σ and n_π of σ and π orbitals respectively come out as integers. The total energy is of the form

$$E = E_\sigma + E_\pi + E_{\sigma\pi} \qquad (3.6)$$

---

[b] The names 'horizontal and vertical correlation' [9] derive from the idea that horizontal correlation allows the electrons to avoid each other on the same side of the molecular plane and vertical correlation has a similar effect perpendicular to the plane.

It is customary to define the π electron energy $\tilde{E}_\pi$ as $E_\pi + E_{\sigma\pi}$; then,

$$E = E_\sigma + \tilde{E}_\pi \tag{3.7}$$

In this partition (which is closely related to what Mulliken called 'cumulative partition' [4]) E_σ is the energy of the σ electrons in the field of the bare nuclei and $\tilde{E}_\pi$ is the energy of the π electrons in the field of the bare nuclei and the σ electrons. Following Lykos and Parr [3], one may regard the σ electrons as the 'core' and the π electrons as the 'peel' and consider the σ–π separation as a case of a 'core-peel' separation. Effective Schrödinger equations where the σ (or π) electrons are represented only through the potential created by their charge distributions can be derived from the expression (3.5) for the π (or σ) system. Therefore, an iterative procedure has to be used. This is the so-called σ–π separation with an 'adjustable' σ core [3]. We shall come back in Sect. 5.1 to the more restrictive assumption of the σ–π separation with a rigid σ core.

Alternative partitions of the whole system into groups according to Eq. (3.5) are possible. One of them is into K shell groups and a valence shell group, which one may call the K-V separation. It is in some respects better justified than the σ–π separation and has been checked by ab initio calculations of small molecules [5,6]. One drawback of the K-V separation is that the strong orthogonality condition is not automatically satisfied for symmetry reasons, as in the σ–π separation. However, this does not lead to serious difficulties.

The essential feature of the σ–π separation is that an effective Hamiltonian can be defined for the π electrons in the field of the nuclei and the σ core. As was pointed out by Sinanoğlu [7], this separation can be derived under conditions more general than the Lykos-Parr assumption. A slightly different formulation of the σ–π separation can be obtained by the methods of second quantization [8].

3.2. Limits of the σ–π Separation

Most organic molecules contain an even number of electrons and have zero spin in the ground state. On the other hand, radicals are systems with an odd number of electrons, and have at least one unpaired electron. The highest energy level (in the sense of the simplest IPM) is supposed to be occupied by the unpaired electron, and in most unsaturated planar radicals the corresponding orbital is a π orbital.

The spin density, i.e. the probability density for finding an unpaired electron spin close to a given nucleus, is responsible for the hyperfine coupling observed in ESR spectra. Now, experimental data show that the magnetic moments of the protons interact with those of the unpaired electron. Consequently, the spin density in the molecular plane (where the H atoms are located) is non-zero. Let us consider a wave function of the form (3.5):

$$\Psi = \mathscr{A} \{^1\Sigma\,(1,2,\ldots,n_\sigma)\,{}^2\Pi\,(n_{\sigma+1},n_{\sigma+2},\ldots,n_\sigma+n_\pi)\}$$

It consists of a singlet Σ function (*i.e.* all σ electrons are paired) and of a doublet Π function (with an odd number of electrons). Therefore, the σ spin density should be zero and only the π spin density should be non-zero. However, the π spin density vanishes in the molecular plane for symmetry reasons, so that there should be a zero total spin density in the plane — and at the position of the nuclei. This is obviously in contradiction with the experimental results and only means that one has to go beyond the approximation of the function (3.5) if one wishes to explain the non-zero spin densities at the position of the protons.

This has been done successfully by several authors (see [10–16]). The essential idea is to replace the simple Lykos-Parr wave function by the following linear combination:

$$\Psi = a_1\mathscr{A}\,\{^1\Sigma\,{}^2\Pi\} + a_2\,{}^2\theta\,\mathscr{A}\,\{^3\Sigma^2\Pi\} \tag{3.8}$$

where $^3\Sigma$ is a triplet function (*i.e.* a function with two unpaired spins) for the σ core and where $^2\theta$ is an operator ensuring that $^3\Sigma$ and $^2\Pi$ are coupled to give a doublet state (with one unpaired electron). It turns out that the coefficient a_2 is only a few percent (in absolute value) of a_1, but the configuration $(^3\Sigma^2\Pi)$ has unpaired electrons on σ orbitals, which are responsible for the hyperfine structure of the ESR spectra. An important result of the theory is that the spin density at the position of a proton is roughly proportional to the probability with which the unpaired π orbital occupies the $2p\pi$ orbital of the adjacent carbon atom (the so-called *McConnell relation*).

For further details, the reader is referred to the original papers. We have made this point to emphasize that a completely correct description of unsaturated molecules is given only by a linear combination of different Lykos-Parr type wave functions. However, one term in this linear combination may have a coefficient far larger than all the others; then, the latter can be neglected unless the contribution of the leading term to a given physical property vanishes or is very small, as is the case for the ESR spectra of unsaturated radicals. In the standard spin polarization interpretation of ESR spectra, these terms are included through singly excited configurations with *three unpaired electrons*: the unpaired π electron of the primitive ground state configuration, and two uncoupled σ electrons, the one on a bonding σ orbital, the other on a σ^* antibonding orbital. The spin polarization mechanism does not change the integral occupation numbers n_σ and n_π of the chemical formula; for that, one ought to include doubly-excited configurations.

It should be added that the so-called *negative spin density* suggested by experimental evidence is explained through excited configurations with three unpaired π electrons, which may be understood as a horizontal spin polarization.

Similar conclusions can be drawn from the study of NMR spectra given by unsaturated closed-shell molecules (see [18-21]). However, the theoretical analysis of the hyperfine structure is more involved for NMR spectra than for ESR spectra, because the nuclear spin-spin coupling constants are second-order phenomena as compared with the electron-nucleus coupling constants.

3.3. Correlation Effects in π Electron Systems

If a n-electron wave function is limited to a Slater determinant of n spin orbitals, one stays within the frame of the independent-particle model, and the best model of that sort (for a discussion, see [22]) for a given problem is that in which the orbitals used to construct the wave function are solutions of the Hartree-Fock equations. This model is only an approximation of the correct wave function. As mentioned in Sect. 3.1, the wave function should be written as a linear combination of Slater determinants, as in Eq. (3.4). To illustrate this, let us consider a two-electron system where the spin can be separated off, so that it is sufficient to consider a function $\tilde{\psi}$ (1,2) depending only on the space coordinates of the two particles 1 and 2. For a singlet state $\tilde{\psi}$ (1,2) is symmetric with respect to space coordinates:

$$\tilde{\psi}(1,2) = \tilde{\psi}(2,1) \tag{3.9}$$

In the frame of the independent particle model, $\tilde{\psi}$ will be a simple product of orbitals

$$\tilde{\psi}(1,2) = [\varphi(1)\,\varphi(2)] \tag{3.10}$$

whereas in a configuration interaction expansion $\tilde{\psi}$ has the form

$$\tilde{\psi}(1,2) = [\sum_{i,j} a_{ij}\varphi_i(1)\,\varphi_j(2)] \quad (a_{ij} = a_{ji}) \tag{3.11}$$

An example of a planar system with two electrons is given by the H_3^+ ion. The space function φ of Eq. (3.10) is a σ orbital and $\tilde{\psi}$ is symmetric with respect to a reflection at the molecular plane. The first π orbital is not used to construct the independent-particle model of the ground state, because it has a much higher energy. However, $\tilde{\psi}$ would be also symmetric if φ were a π orbital, because in a reflection at the molecular plane both φ (1) and φ (2) would change signs, so that there is no overall change. Therefore, the space functions φ_i of the more general expression (3.11) may be σ or π orbitals, but the coefficients a_{ij} are different from zero only if φ_i and φ_j belong to the same symmetry species (σ or π). The energy given by the function (3.10) for the ground state of H_3^+ is -1.2971 a.u. and the energy lowering obtained by using eight functions φ_i in (3.11) is equal to 0.0388 a.u. [23]; in this figure, 0.0340 a.u. originates from a configuration interaction among σ orbitals and only 0.0048 a.u. from the interaction with the first two π orbitals. The corresponding occupation numbers are $n_\sigma = 1.9971$ and $n_\pi = 0.0029$. By extrapolation, one finds for the total energy of H_3^+ $E = -1.336$ a.u. if σ-type orbitals only are used in Eq. (3.11) (the so-called 'σ limit') or $E = -1.342$ a.u. if π orbitals are included.

Now, if one analyzes the correlation of the positions of the two electrons brought about by the configuration interaction treatment of H_3^+, an important result becomes apparent (see e.g. [24]). In the independent particle model, the position coordinates x_1, y_1, z_1 of one electron and those x_2, y_2, z_2 of the other electron are completely independent of each other: there is no correlation. From the configuration interaction function constructed from σ orbitals only, one finds that the two x coordinates, which are perpendicular to the molecular plane, are still independent, whereas the probability for y_1 to be close to y_2 and z_1 to be close to z_2 is considerably reduced; this result can be considered as an effect of 'horizontal' correlation (the electrons try to avoid each other horizontally). A CI function constructed from the lowest SCF σ orbital and from additional π orbitals leads to a picture where the y and z coordinates are independent, but where there is little probability of close values of x_1 and x_2; in other words, such a function accounts for 'vertical' correlation (the electrons tend to be on opposite sides of the plane). A correct wave function would allow for both horizontal and vertical correlation. It is easy to understand why the independent-particle model is unsatisfactory. Because of their identical charges the electrons repel each other; their average interaction, i.e. the interaction of the corresponding charge clouds, is taken into account even in the IPM, but the IPM ignores the fact that, owing to this repulsion, the electrons tend to occupy different places. A CI function allows them to do so; this is why the energy of a CI expansion is lower than that of a single Slater determinant. However, the 'correlation energy' is a small correction to the interaction energy. The situation is illustrated in Table 3, where some properties of diatomic molecules as calculated from the molecular Hartree-Fock equations are compared with the exact values. Except for F_2, correlation has not much effect on equilibrium distances and force constants, but binding energies calculated from Hartree-Fock theory are very poor. Similarly, the results for spectral transition energies are not in good agreement with experiment.

Table 3. *SCF and multi-configuration SCF calculations of observables in diatomic molecules*

References		H_2 a)	Li_2 a)	N_2 b)	F_2 a)	LiH c)	HF c)
Binding	SCF	3.64	0.17	5.27	−1.37	1.49	4.38
energies -D_e	MC-SCF[2]	4.63	0.99		0.95		
in eV	Exp	4.75	1.05	9.90	1.68	2.52	6.12
Equilibrium	SCF	0.73	2.78	1.06	1.32	1.60	0.90
distances R_e	MC-SCF	0.74	2.69		1.43		
in Å	Exp.	0.74	2.67	1.10	1.42	1.60	0.92
Vibration	SCF	4561	326	2729	1257	1433	4469
frequencies ω_e	MC-SCF	4398	345		750		
in cm^{-1}	Exp	4400	351	2358	892	1405	4139

[1] SCF calculations with large optimized basis of Slater orbitals.
[2] SCF culculations with multi-determinant wave functions including double excitations from valence molecular orbitals.
[a] Das, G., Wahl, A. C.: J. Chem. Phys. *47*, 2934 (1967).
[b] Cade, P. E., Sales, K. D., Wahl, A. C.: J. Chem. Phys. *44*, 1973 (1966).
[c] Cade, P. E., Huo, W. M.: J. Chem. Phys. *47*, 614 (1967).

Electron correlation probably also plays an important role in the theory of unsaturated or *conjugated organic molecules*. Unfortunately, accurate numerical data are so far lacking, and we must discuss simplified models in order to understand the electron correlation in π electron systems. Although different contributions to electron correlation should be considered, namely the horizontal and the vertical correlation of the σ and π systems and the intergroup correlation, so far the correlation of the π system alone has been most studied. As has been mentioned, the horizontal correlation of the π system can be taken into account within the framework of the $\sigma-\pi$ separation.

If one confines the study of a molecule to its π electrons, the most general function for its π system is given by Eq. (3.5). To calculate Π, it is necessary first to make plausible assumptions about the potential created by the charge distribution of the σ system (see Sect. 5.1) and then to choose a basis set of π orbitals and perform a configuration interaction for the π system. The simplest basis is that of the atomic $2p\pi$ orbitals of the constituent atoms. If one expresses $\varPi$ as a linear combination of all the Slater determinants that can be constructed from those atomic orbitals, this treatment may be referred to as 'full' configuration interaction [25,26,27] the term 'complete' being reserved for the case where the orbital basis is complete[c].

It is well known that, in order to obtain reasonable predictions for spectra, a CI treatment is preferable to the simple one-particle excitation model (see Sect. 5.3). However, allowing for horizontal correlation through CI with π orbitals is not sufficient to restore agreement with experiment (for a discussion, see [28]). We come back to this point at the end of this section.

Full CI is rather complicated, particularly for large molecules. Therefore, many approaches have been suggested with the aim of getting the same (or almost the same) result as from „full CI", but in an easier way. Among these works, one has to mention especially the '*alternant molecular orbital method*' (AMO) (see *e.g.* [29]), the '*non-paired spin orbital method*' (NPSO) [30], various electron pair treatments [31] and, finally, rather sophisticated schemes borrowed from solid state or quantum field theory [32,33]. All these approaches are concerned less with the actual physical problem of the horizontal correlation of the π electron system than with the general correlation problem. The π systems just furnish well-defined models, for which by definition a full CI calculation gives the exact solution, to be compared with various simplified treatments. For this reason, we shall not discuss these approaches in detail here. Suffice it to mention the following result concerning the horizontal correlation of π electrons: in systems like the *polyenes*, where the alternation of bonds suggests a description in terms of localized π bonds (see Sect. 4.5), almost the entire horizontal correlation energy is due to the 'intrapair' correlation energy of the electron pairs localized on the double bonds [31]. This means that the function for the π group can be written to a good approximation as a product of pair functions

$$\varPi\ (1,2,\ldots,n_\pi) = \varPsi_A(1,2)\varPsi_B(3,4)\ldots\varPsi(n_{\pi-1},n_\pi) \qquad (3.12)$$

The situation is more complicated for systems with typically delocalized orbitals like benzene [33,34,35], where the total horizontal correlation energy cannot be broken down into contributions from different electron pairs.

---

[c] The term '*full CI*' was introduced by de Heer in connection with the problem of benzene [25].

Vertical correlation is more difficult to treat than horizontal correlation because it goes beyond the σ—π separation. Actually, the π wave function should be completed by Slater determinants partially constructed from σ orbitals, which should be orthogonal to the whole set of occupied σ orbitals. Calculations of this sort are rather complicated and almost nothing has been done on these lines, apart from some calculation of valence states of carbon to be discussed below. Most approaches to the vertical correlation of π electron systems have been of a semi-empirical nature. An example of such an approach is the method 'Atoms in Molecules' [36] which has been applied to a number of π electron systems. It was recognized that the atomic correlation errors have to be corrected in order to get reliable results for molecules [37,38] (see Sect. 5.3).

Pariser and Parr [39,40] proposed an amendment to the π electron theory which was justified on somewhat similar lines. Here, we are concerned only with one aspect of the *Pariser-Parr theory*, namely the reduction of the Coulomb repulsion energies of electrons. In order to reproduce the spectra of π electron systems, one is forced (among other factors) to reduce the 'one-center electron repulsion integrals' of the carbon atom

$$\int \chi_p^*(\mu)\, \chi_p(\mu)\, \frac{1}{r_{\mu\nu}}\, \chi_p^*(\nu)\, \chi_p(\nu)\, d\tau_\mu\, d\tau_\nu \tag{3.13}$$

to about 10 *eV*, whereas the theoretical value calculated by taking for the orbital χ_p a Slater $2p\pi$ orbital is 17 *eV*. Such a reduction is usually justified by the following argument: consider the '*disproportionation reaction*'

$$C + C \longrightarrow C^+ + C^- \tag{3.14}$$

where each of these carbon atoms is in its trigonal valence state. The energy change in this reaction is given experimentally by the difference *I-A* of the ionization potential *I* and the electron affinity *A* of carbon in its valence state and amounts to about 10 *eV*. The theoretical energy difference is given by the integral (3.13), provided that one assumes there is no change in either the σ core or the $2p\pi$ orbitals when passing from the neutral atom to its ions. Since these two assumptions are not realistic, the argument is not fully convincing, but it is probable that the theoretical value for the integral (3.13) *overestimates the actual repulsion energy* of two $2p\pi$ electrons at the same atom. The current explanation is that one has to use a smaller value in order to account for correlation effects. In fact, if one wants to account for electron correlation while keeping the formalism of the independent particle model, *i.e.* if one wants to avoid a CI treatment, one can note that correlation does not much change the expectation values of typical one-particle operators, like the kinetic energy or the potential energy in the field of the core. It does however, change appreciably the electron repulsion energy. So, if one disregards the question whether a theory allowing for correlation within an independent particle scheme can be formulated in a consistent way (see *e.g.* [41]) and agrees to treat an effect going beyond the $\sigma-\pi$ separation within the formalism of this separation, it is plausible that one should take smaller values for the two-electron integrals.

There are in the literature several attempts to introduce correlation into π electron theories in a systematic way, *e.g.* the correlation factor methods of Kołos [42] and Julg [43], and the *split p-orbital method* (SPO) of Dewar [44,45]. When applied to the integral (3.13), all these methods give about the same value, and this is the main reason why they work for the calculation of spectra. The SPO method has been strongly criticized on mathematical grounds [46,47,48]; however, it has the merit of having clarified the fact that the type of correlation to be introduced through a reduction of the electron repulsion integrals is vertical correlation, whereas horizontal correlation should be accounted for through full CI in the π AO's basis. Since this vertical correlation is also present in atoms (*e.g.* in C^-), it is not astonishing that it can be estimated from atomic term values (*e.g.* based on the reaction (3.14)).

The argument based on correlation is perhaps convincing from a qualitative point of view, but it is very difficult to interpret a difference of 7 *eV* as the correlation energy of a pair of $2p\pi$ electrons. *Ab initio* calculations of simple systems [49,50,51] lead to estimates of at most 1 *eV* for the correlation energy of one pair of $2p\pi$ electrons. Approximate calculations of the vertical correlation energy in the C^- valence state [52,53] lead to similar values. The argument that the electron repulsion energy should be reduced by twice the correlation energy, because the kinetic energy is increased by the amount of the correlation energy [41,54] can hardly explain a lowering of 7 *eV*. In conclusion, the greater part of the difference between the 'theoretical' and 'semi-empirical' values of the one-center electron repulsion integral is not due to correlation and has to be explained in an alternative way.

Actually, the integral (3.13) is somewhat reduced by using for the atomic orbital $2p\pi$ of carbon more elaborate forms than a simple Slater function, for instance Hartree-Fock orbitals [54]. Table 4 gives some values obtained in this way [55].

The electron repulsion integral between two charge distributions $\chi_p^*(1)\,\chi_p(1)$ and $\chi_q^*(2)\,\chi_q(2)$ located *on two different atoms* is still more reduced than the one-center integral; at the same time, the corresponding overlap integral S_{pq} increases from 0.25 for two Slater functions at 1.4 A apart to 0.33 for two Hartree-Fock functions [56]. This shows the importance of an appropriate choice for the basis set of atomic orbitals used in π calculations, especially for electronic transitions which depend on the difference between repulsion integrals rather than their absolute values [57].

Table 4. *Electron repulsion integrals (in eV) the $2p\pi$ orbitals of carbon*

	One-center integrals (in eV)	Two-center integrals at the distance $R = 3.25$ Å (in eV)
Slater function ($Z_c^* = 3.18$)	16.93	4.42
Double-zeta function[1]	15.74	1.02
Hartree-Fock functions[2]	15.25	
Experimental value[3]	11.13	

[1] Calculated from Clementi's tables: Clementi, E.: IBM J. Res. Dev. *9*, 2 (Suppl.) (1965).

[2] Calculated from analytical SCF orbitals: Arai, T., Lykos, P.: J. Chem. Phys. *38*, 1447 (1963).

[3] Calculated using ionization potential and electroaffinity of carbon in the valence state V_4, $tr^3\pi$: Hinze, J., Jaffé, H. H.: J. Am. Chem. Soc. *24*, 540 (1962).

It has been noted by several authors [54,58,59,60] that the hypothetical charge transfer reaction (3.14) has to be treated in a more complicated way; among other things, the effective nuclear charge of the carbon atom in its neutral valence state should be rather different from the charge of its positive or negative ion. In fact, if one uses AO's minimizing the energy of appropriate valence states [61], the discrepancy between the theoretical and empirical energy of the hypothetical reaction (3.14) is very much reduced. The one-center electron repulsion integral (3.13) for carbon turns out to be about 12.4 *eV*. It may be that the remaining difference with respect to the 10 *eV* of Pariser and Parr is due to vertical correlation energy.

It is outside the scope of the present review to discuss further electron correlation in the case of π electrons. As regards correlation effects in σ systems, they should be more easily understandable, at least if the σ system consists of localized bonds; however, very little is actually known about them. An example where intergroup correlation effects are important has been discussed in Sect. 3.2.

3.4. References

[1] McWeeny, R.: Proc. Roy. Soc. (London) *A 223*, 306 (1954); Rev. Mod. Phys. *32*, 335 (1960).
[2] Parr, R. G., Ellison, F. O., Lykos, P. G.: J. Chem. Phys. *24*, 1106 (1956).
[3] Lykos, P. G., Parr, R. G.: J. Chem. Phys. *24*, 1166 (1956).
[4] Mulliken, R. S.: J. Chim. Phys. *46*, 497, 675 (1949).
[5] Ahlrichs, R., Kutzelnigg, W.: J. Chem. Phys. *48*, 1819 (1968).
[6] Bender, F., Davidson, E. R.: J. Phys. Chem. *70*, 2675 (1966).
[7] Sinanoĝlu, O.: J. Chem. Phys. *36*, 3198 (1962).
[8] Harris, R.: J. Chem. Phys. *47*, 3967 (1967).
[9] Dewar, M. J. S.: Rev. Mod. Phys. *35*, 586 (1963).
[10] McConnell, H. M., Chesnut, D. B.: J. Chem. Phys. *28*, 107 (1958).
[11] Lefebvre, R., Dearman, H. M., McConnell, H. M.: J. Chem. Phys. *32*, 176 (1960).
[12] McLachlan, A. D., Dearman, H. H., Lefebvre, R.: J. Chem. Phys. *33*, 65 (1960).
[13] Colpa, J. P., Bolton, J. R.: Mol. Phys. *6*, 273 (1963).
[14] Giacometti, G., Nordio, P. L., Pavan, M. V.: Theoret. Chim. Acta *1*, 404 (1963).
[15] McWeeny, R., Sutcliffe, B. T.: Mol. Phys. *6*, 493 (1963).
[16] Berthier, G., Lemaire, H., Rassat, A., Veillard, A.: Theoret. Chim. Acta *3*, 213 (1965).
[17] Malrieu, J. P.: J. Chem. Phys. *46*, 1654 (1967).
[18] Acrivos, S. V.: Mol. Phys. *5*, 11 (1962).
[19] Karplus, M.: J. Chem. Phys. *33*, 1847 (1960); *50*, 3133 (1969).
[20] Barbier, C., Berthier, G.: Int. J. Quant. Chem. *1*, 657 (1967).
[21] Ditchfield, R., Murrell, J. N.: Mol. Phys. *15*, 533 (1968).
[22] Kutzelnigg, W., Smith, V. H.: J. Chem. Phys. *41*, 896 (1964).
[23] Kutzelnigg, W., Ahlrichs, R., Labib-Iskander, I., Bingel, W. A.: Chem. Phys. Letters *1*, 447 (1967).
[24] Kutzelnigg, W., Del Re, G., Berthier, G.: Phys. Rev. *172*, 49 (1968).
[25] de Heer, J.: Rev. Mod. Phys. *35*, 631 (1963).
[26] Koutecky, J., Hlavaty, K., Hochmann, P.: Theoret. Chim. Acta *3*, 341 (1965).
[27] Moskowitz, J. W., Barnett, M. P.: J. Chem. Phys. *39*, 1557 (1963).
[28] Parr, R. G.: Quantum Theory of Molecular Electronic Structure. New York: Benjamin 1963.
[29] Pauncz, R.: Alternant Molecular Orbital Method, Philadelphia: Saunders 1967.
[30] Empedocles, D. D., Linnett, J. W.: Theoret. Chim. Acta *4*, 377 (1966); Proc. Roy. Soc. (London) *A 282*, 166 (1964).
[31] Staemmler, V., Kutzelnigg, W.: Theoret. Chim. Acta *9*, 67 (1967).
[32] Linderberg, J., Öhrn, Y.: Proc. Roy. Soc. (London) *A 285*, 445 (1965). — Öhrn, Y., Linderberg, J.: Phys. Rev. *139*, A 1063 (1965).
[33] Čížek, J.: J. Chem. Phys. *45*, 4256 (1966).
[34] Staemmler, V., Kutzelnigg, W.: (to be published).
[35] Ebbing, D. D., Poplawski, L. E.: J. Chem. Phys. *45*, 2657 (1966).
[36] Moffitt, W.: Proc. Roy. Soc. (London) *A 210*, 245 (1951).
[37] Hurley, A. C.: Proc. Phys. Soc. (London), *68*, 149 (1955); *69*, 301 (1956).
[38] Arai, T.: Rev. Mod. Phys. *32*, 370 (1960).
[39] Pariser, R.: J. Chem. Phys. *21*, 568 (1953).
[40] Pariser, R., Parr, R. G.: J. Chem. Phys. *21*, 466, 767 (1959).
[41] Kutzelnigg, W.: Theoret. Chim. Acta *1*, 327 (1963).
[42] Kołos, W.: Acta Phys. Pol. *16*, 257, 299 (1957).
[43] Julg, A.: Theoret. Chim. Acta *2*, 134 (1964).
[44] Dewar, M. J. S., Hojvat, N. L.: J. Chem. Phys. *34*, 1232 (1961).

45) Dewar, M. J. S., Sabelli, N. L.: J. Phys. Chem. *66*, 2310 (1962).
46) Griffith, J. S.: J. Chem. Phys. *36*, 1689 (1962).
47) Coulson, C. A., Sharma, C. S.: Proc. Roy. Soc. (London) *A 272*, 1 (1963).
48) Saturno, A. F., Joy, H. W., Snyder, L. C.: J. Chem. Phys. *38*, 2494 (1963).
49) Nesbet, R. K.: Phys. Rev. *155*, 56 (1967).
50) Ahlrichs, R., Kutzelnigg, W.: Chem. Phys. Letters *1*, 651 (1968).
51) Bender, C. F., Davidson, R.: Phys. Rev. *183*, 23 (1969).
52) Hermann, R. B.: J. Chem. Phys. *42*, 1027 (1965).
53) Lowe, J. P.: J. Chem. Phys. *41*, 502 (1964).
54) Arai, T., Lykos, P. G.: J. Chem. Phys. *38*, 1447 (1963).
55) Ažman, A., Zakrajšek, E.: Z. Naturforsch. *23a*, 440 (1968).
56) Mulliken, R. S., Rieke, C. A., Orloff, D., Orloff, H.: J. Chem. Phys. *17*, 1248 (1949).
57) Lykos, P. G.: J. Chem. Phys. *35*, 1249 (1961).
58) Ellison, F. O.: J. Chem. Phys. *37*, 1414 (1962).
59) Ellison, F. O., Huff, N. T.: J. Chem. Phys. *38*, 2444 (1963).
60) Orloff, M. K., Sinanoğlu, O.: J. Chem. Phys. *43*, 49 (1965).
61) Silverstone, H. J., Joy, H. W.: J. Chem. Phys. *47*, 1384 (1967).

4. Specific properties of σ and π Electrons

4.1. Some Current Statements Concerning σ and π Electrons

In Chapt. 2 we have recalled how σ and π orbitals can be defined in terms of a rigorous theory and what the notions of σ and π electrons actually mean. In Chapt. 3 we have introduced the $\sigma-\pi$ separation and discussed its justification and limitations. If the $\sigma-\pi$ separation is valid, then an effective Hamiltonian for the π electrons can be constructed into which the σ electrons enter only via the effective potential created by their charge distribution.

Of course, the basic question is: why stress the distinction between σ and π orbitals and the $\sigma-\pi$ separation? Is this point of view really useful, or is it just a trivial by-product of the quantum-mechanical treatment? A partial answer has already been given in Chapt. 1: to some extent, the notion that there are two classes of electrons associated with quite different molecular properties is suggested by experimental evidence These two classes have been identified with the σ and π electrons defined in the preceding sections, and this has resulted in a number of con-clusions regarding both the 'theoretical' and 'experimental' differences between σ and π electrons. These conclusions can be summarized in the following rather familiar statements:

a) σ and π electrons are localized in different regions of space, the σ electrons more on the plane, the π electrons above and beneath the plane;

b) π electrons are more loosely bound and more easily polarizable than σ electrons;

c) σ electrons form localized bonds, π electrons are delocalized;

d) π electrons are chemically more reactive than σ electrons.

None of these statements is rigorously true, but all hold *grosso modo*. In the following sections, we shall comment on them in more detail.

4.2. Spatial Distributions

Statement (a) is based on the fact that π orbitals have a nodal plane in the plane of the molecule (or the unsaturated or conjugated part of it)

Orbital
1a_g
(K-shell)

a

Orbital
1b_{3u}
(K-shell)

b

Orbital
2a_g
(σ)

c

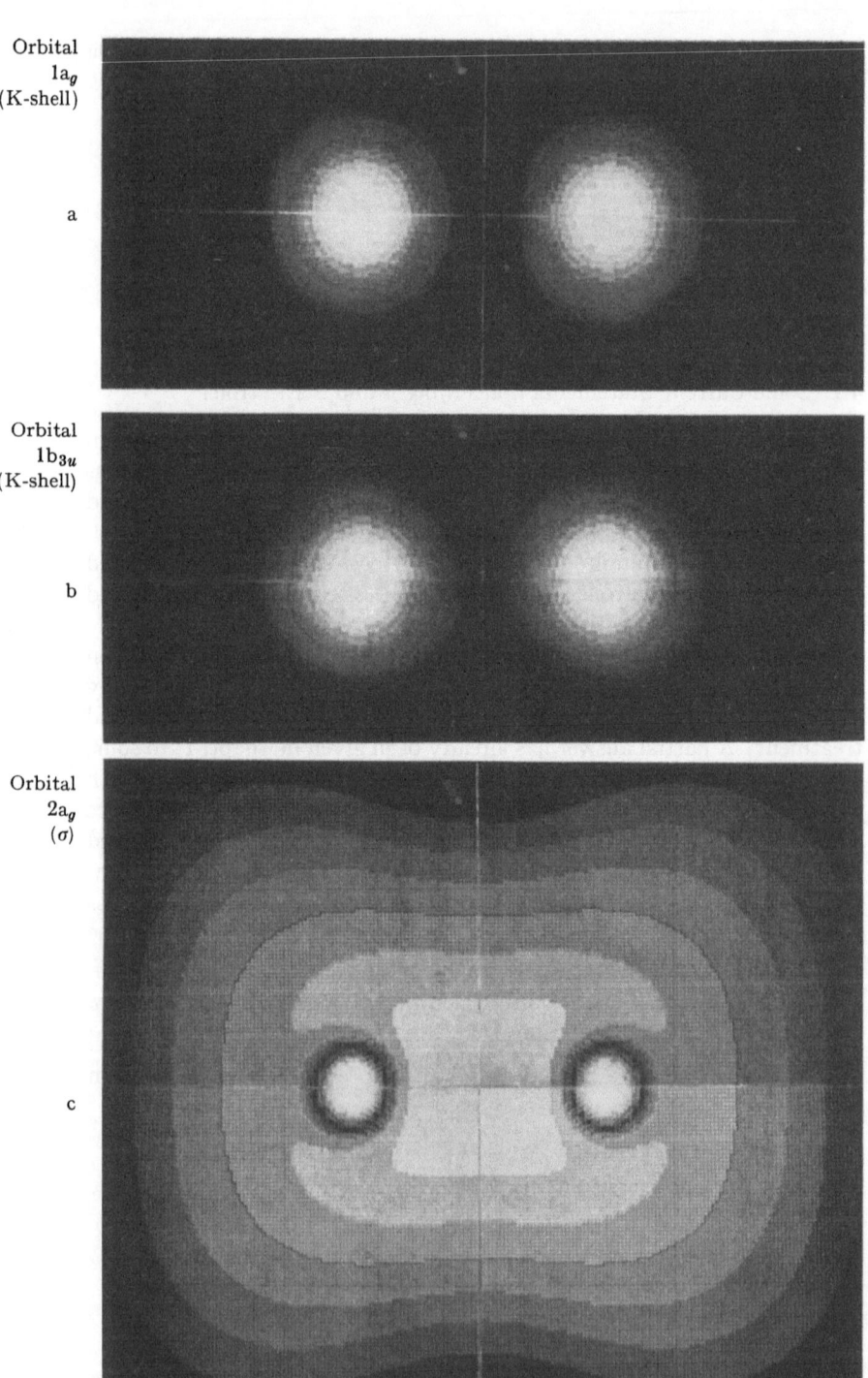

Fig. 2a—c

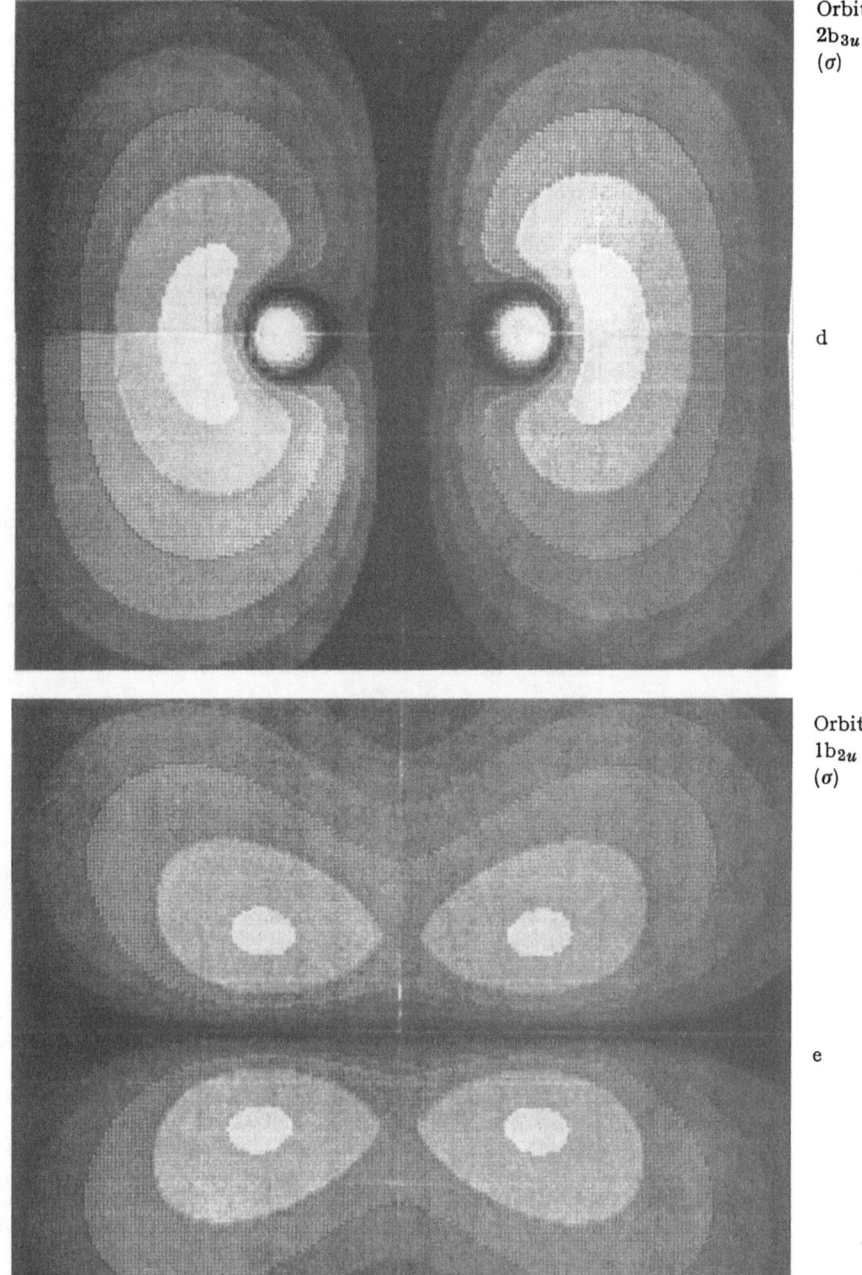

Orbital
2b_{3u}
(σ)

d

Orbital
1b_{2u}
(σ)

e

Fig. 2 d and e

Fig. 2 a—h. Orbital densities in the ethylene molecule. (For 1b_{1u} in a plane perpendicular to the molecular plane through the C—C axis for the other MO's in
the molecular plane). After H. Preuss, private communication.

Orbital
3a$_g$
(σ)

f

Orbital
1b$_{1g}$
(σ)

g

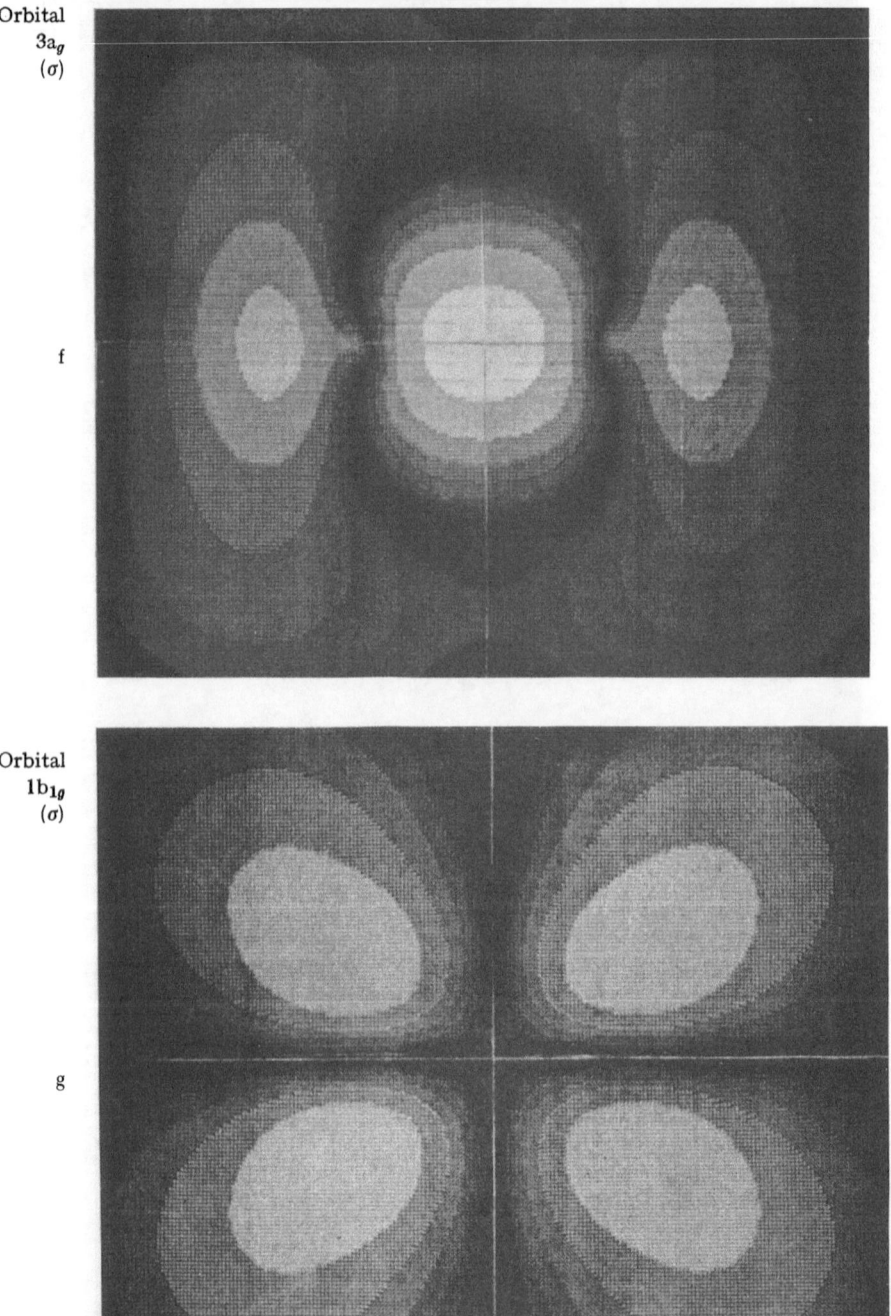

Fig. 2 f and g

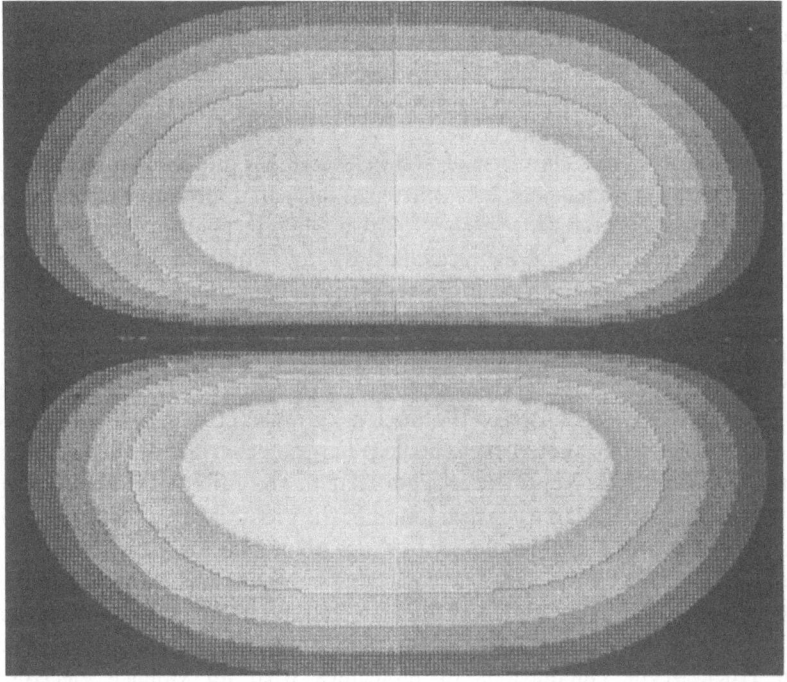

Fig. 2h. Orbital $1b_{1u}$ (π)

and that the probability of finding a π electron in the molecular plane vanishes. On the other hand, the σ electron density has its maximum in the molecular plane. The situation is best illustrated by the orbital density maps of Fig. 2.

If the π electrons were strictly *outside* the σ electron cloud, the potential created by the π charge distribution at the position of the σ electrons would practically vanish, at least in non-polar molecules. This results from the fact that the repulsive electrostatic potential due to one lobe of the π charge distribution is nearly cancelled by the potential of the other lobe, except at the end of the molecule. Then the effective Hamiltonian governing the motion of the σ electrons (in the frame of the $\sigma-\pi$ separation) would be practically the same as that of the ion in which all the π electrons are ionized away. The presence of the π electrons would be felt very little by the σ electrons, except in systems with highly polar σ bonds. Then, an iterative procedure adjusting successively the σ charges to the π charges and *vice versa* would not be necessary, and one and the same σ core could be used for different π states of the molecule. In this idealized description, the potential created by the σ electrons would be very close to that of point charges at the nuclear positions, the effect of the σ electrons on the π electrons being essentially to 'shield' the real nuclear charges.

Actually, the assumption of non-interacting σ and π charges is much too crude a simplification. As a matter of fact, an approximate treatment of benzene suggested that there is appreciable interpenetration of σ and π densities [1], and recent SCF calculations on simple molecules support this idea [2]. It is known from Slater's rules [3] that 2s and 2p electrons, say in the carbon atom, shield the nuclear charge from the other electrons in the same shell to about the same extent, namely by about 0.35 units. Now, if 2s and 2p electrons were localized in different regions of space, the 2s charge being closer to the nucleus, a much stronger shielding due to the 2s electrons should be expected. On the other hand, the situation in a molecule may be different from that in the isolated atoms; for instance, the best effective charges found in the ground state of acetylene [4], hydrogen cyanide or formaldehyde [5] are about 3.5 for a 2s AO, 4.0 for a $2p\sigma$ AO and 3.0 for a $2p\pi$ AO of carbon; in other words, the π electrons see a somewhat more shielded nuclear charge than the σ electrons do. It should be noted that interpenetrating σ and π densities are also found in Hückel-type calculations of molecules with all valence electrons (see Sect. 6.2); therefore, this result is not basically a σ–π interaction effect, since the extended Hückel method completely ignores the Coulomb repulsion of electrons.

Briefly, it may be stated that *there is a large overlap* between σ and π densities, but the σ cloud is closer to the molecular plane than the π cloud and that the influence of the σ electrons on the π electrons is more pronounced than the reverse effect.

4.3. Bonding Properties

Statement (b) is based on the remark that the 2p orbitals of atoms are more loosely bound than 2s orbitals as is reflected by the corresponding ionization potentials. Now, in the LCAO approximation, π orbitals are constructed from atomic $2p_x$ orbitals only, whereas σ bonds involve 2s, $2p_y$ and $2p_z$ orbitals to roughly the same extent. In these considerations it is convenient to assume that the atoms are in appropriate valence states [6,7] and to take the corresponding ionization potentials [8,9]. In the case of carbon, for example, one should consider the ionization potential of a trigonal hybrid and a $2p\pi$ electron in the $(V_4, tr^3\pi)$ valence state (Table 5). The ionization potential of a hydrogen 1s orbital is also indicated, because this orbital is involved in σ binding.

Table 5. *Ionization potentials of atoms in valence states*

Atomic states	Promotion energies (in eV)		Ionization potentials (b) (in eV) of Valence electrons or lone pairs		
	SCF[1]	Exp.[2]	s	p	s^2 or p^2
Carbon					
C $^3P\,s^2p^2$	0	0	19.5	10.7	—
C $V_2\,s\,p^2p$	10.72	9.83	21.42	11.68	p^2 9.92
C $V_4\,s\,p\,p\,p$	8.76	8.48	21.01	11.27	—
Nitrogen					
N $^4S\,s^2p^3$	0	0	25.6	12.9	—
N $V_3\,s^2p\,p\,p$	1.90	1.08	—	13.94	s^2 25.58
N $V_3\,s\,p^2p\,p$	15.92	14.29	26.92	14.42	p^2 12.37
Oxygen					
O $^3P\,s^2p^4$	0	0	32.4	15.9	—
O $V_2\,s^2p^2p\,p$	0.54	0.54	—	17.28	$\left\{ \begin{array}{l} s^2 \;\; 32.30 \\ p^2 \;\; 14.61 \end{array} \right.$
Hydrogen					
H $^1S\,s$			13.60	—	—

[1] Hartree-Fock calculations of atomic valence states: Kochanski, E., Berthier, G.: In: Structure hyperfine magnétique des atomes et des molécules, 177, Paris: C.N.R.S. 1967.

[2] Experimental data for L-shell ionization potentials: Slater, J. C.: In: Quantum Theory of Atomic Structure, Vol. I, p. 206. New York: McGraw 1960, and valence state ionization potentials: Hinze, J., Jaffé, H. H.: J. Am. Chem. Soc. *84*, 540 (1962).

Of course, the energy levels of the molecular orbitals are not identical with those of the AO's from which they are constructed. To a first approximation, one may say that the *AO levels are split into bonding and antibonding (and possibly non-bonding) levels*. Since σ bonding is stronger the splitting of the σ levels is supposed to be larger and the bonding and antibonding π levels should be closer to the zero-level than the corresponding σ levels. Actually, a number of bonding σ levels lie lower than bonding π levels, but there is no reason why all of them should lie below the bonding π levels. The actual situation is schematized in Fig. 3.

In reality, all-electron calculations of the *benzene molecule* give the result that (at least) one occupied σ orbital has a higher orbital energy than the lowest π level (see Table 14.) If one analyzes cases of this sort, one finds that the high-lying σ levels belong to C—H rather than C—C

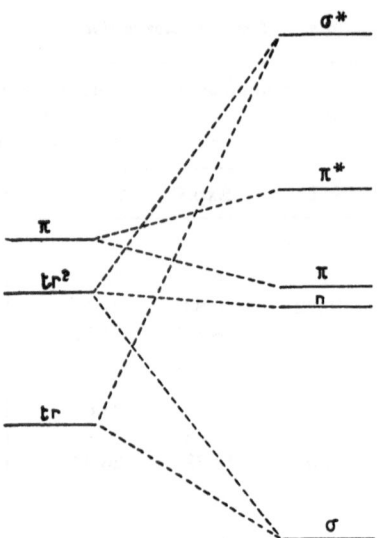

Fig. 3. Bonding, non-bonding, and antibonding molecular orbitals

bonds. In *graphite*, where there are no C—H but only C—C bonds, there is evidence that the σ levels form a completely filled 'valence band' while the π electrons are in a half-filled 'conduction band'. It was believed for a while that the orbital levels of *smaller hydrocarbons* could also be grouped into two 'bands' in a similar way. This is obviously not the case. In addition to the electrons of the C—H bonds, the electrons of nitrogen lone pairs in *aza-compounds* or oxygen lone pairs of *carbonyl compounds* also have relatively small ionization energies; they are essentially non-bonding electrons occupying molecular orbitals whose energies are close to those of bonding π electrons, a fact that is related to the spectral behaviour of molecules containing atoms with lone pairs (see Sect. 5.3).

Contrary to formerly accepted opinion, the lack of any separation of the σ and π levels into two bands with a gap between them has nothing to do with the question of the $\sigma-\pi$ separation as discussed in Sect. 3.1. In fact, all the calculations mentioned in this context were made in the framework of the independent-particle model and *a fortiori* in the frame of $\sigma-\pi$ separation.

Finally, we remark that unlike σ orbitals, π orbitals are much less important as regards the *stability* of a molecule. For instance, one can remove one or two electrons from condensed aromatic systems and obtain very stable ions. Similarly, electrons may be added to antibonding π orbitals without much affecting the stability of the compound. What is true for ionization holds also for excitation: as a rule, excitation of a π electron does not much affect the stability of the molecule (although it may lead to a change of geometry); σ excitation often leads to dissociation.

4.4. Polarizabilities

If a molecule is placed in a homogeneous electric field, its energy changes as a result of orientation and polarization. The new energy E can be expressed as a power series of the field strength $\mathscr{E}$:

$$E = E_0 - m\,\mathscr{E} - \tfrac{1}{2}\,\alpha\,\mathscr{E}^2 \qquad (4.1)$$

where m is the component of the permanent electric dipole moment $\vec{m}$ of the given molecule along the direction of the field, and α is a quantity associated with the induced electric moment and called the *'polarizability'* of the molecule. In general, the induced moment is not parallel to the inducing field and the polarizability is not the same in different directions; therefore, Eq. (4.1) should be written in terms of a symmetric tensor $\boldsymbol{a}$:

$$E = E_0 - (\vec{m}\cdot\vec{\mathscr{E}}) - \tfrac{1}{2}(\vec{\mathscr{E}}\,\boldsymbol{a}\,\vec{\mathscr{E}}) \qquad (4.2)$$

where the second-order term is a doubly contracted product. For a highly symmetric molecule like methane, the components along its principal axes are equal, and Eq. (4.2) reduces to Eq. (4.1). For planar molecules, two of the principal axes of $\boldsymbol{a}$ lie in the molecular plane, the third axis being perpendicular to it, and the 'horizontal' and 'vertical' polarizabilities are in general very different. From the point of view of quantum chemistry, polarizabilities can be expressed as *sums of contributions of individual orbitals*. Consequently, in planar molecules σ and π contributions α_σ and α_π can be defined; both types have in general horizontal and vertical components.

It has been found empirically that the polarizability of a non-conjugated molecule can be decomposed into contributions of different bonds. From Denbigh's [10] analysis of experimental data, one can conclude that the mean polarizability (*i.e.* the average over the three directions) for a $C-H$ bond is about 6.10^{-25} cm^3, that of $C-C$ σ bond 5.10^{-25} cm^3 and that of a localized π bond 8.10^{-25} cm^3.

Values for the various polarizability components have been computed theoretically only in simple molecules, like H_2 [11,12], HF or CO [13,14,15] and the experimental bond polarizabilities presented for hydrocarbons (see *e.g.* [16]) have to be used with caution, because they are usually derived by assuming that the σ contributions are the same in unsaturated and saturated molecules and the C—C bond distances play no role. Actually, the mean values of polarizabilities are reliable to about 20% for usual bonds, but the separate values $\alpha_{\parallel}$ and $\alpha_{\perp}$ assigned to the polarizability in the direction of the bond and the two directions perpendicular to it are more doubtful. In general, parallel polarizabilities are larger than perpendicular ones.

The theoretical expression of α involves an infinite sum over excited states. However, the polarizability of an electron in an orbital φ can be written in the following approximate from

$$\alpha_{xx}(\varphi) \sim \frac{\int \varphi^* x^2 \varphi \, d\tau}{I(\varphi)} \tag{4.3}$$

(and similar expression for α_{yy}; α_{zz})

where $I(\varphi)$ is the ionization potential of the electron under consideration [17]. It follows that the contributions of inner-shell electrons to the polarizability should be negligible, because the numerator is much smaller and the denominator much larger than for valence electrons. This formula also suggests that π electrons should be more easily polarized than σ electrons (especially in the direction perpendicular to the molecular plane), because their ionization potentials are smaller than those of σ electrons and the mean value associated to the square of the x coordinate is larger.

The effects of delocalization on the π electron contribution to the polarizabilities of *conjugated molecules* has been studied by the Hückel method [18,19]. The horizontal polarizability, say, in the direction of the z axis, is related to the Coulson Longuet-Higgins atom-atom polarizabilities $\pi_{r,s}$ by an expression of the form

$$\alpha_{zz} = e^2 \sum_{r,s} z_r z_s \pi_{rs} \tag{4.4}$$

where z_r and z_s are the z coordinates of the rth and s^{th} atoms and the summation is taken over all the pairs of conjugated atoms. The polarizability $\pi_{r,s}$ is the derivative of the π charge on the rth atom with respect to the sth diagonal element of the effective Hamiltonian, or alternatively, the second cross-derivative of the total energy with respect to the rth diagonal elements [20]. In this theory, the perpendicular polarizabilities vanish. The contribution taken into account by Eq. (4.4) comes only from the π charge displacements induced by the electric field and not from the deformation of the atomic orbitals. In order to compare with experiment, one has to add the π polarizabilities obtained for each bond to the contribution of σ bonds[a].

Similar expressions are obtained by perturbation calculations within the frame of the SCF theory of π electron systems (see Sect. 5.1): in aromatic hydrocarbons, the π electrons seem to be responsible for about one-half of the in-plane electric polarizabilities, and their contribution increases with the size of the molecule [21]. The same kind of developments can be made for magnetic susceptibilities [22,23].

On the whole, the analysis of various polarizabilities suggests that the π *electrons are more strongly affected by external perturbations* than the σ electrons. Nevertheless, the most important point is perhaps the low polarizability of σ electrons rather the high polarizability of π electrons, which permits us to regard the σ distribution as a comparatively rigid one.

4.5. Localization in σ and π Bond Systems.

Statement (c) refers to the most significant difference between σ and π bonds; however, it is not correct to say that σ bonds are always localized and π bonds always delocalized. In order to assess the difference clearly, we have to discuss at some length what the terms 'localized' and 'delocalized' really mean.

The molecular orbitals which are solutions of the standard Hartree-Fock equations are delocalized, *i.e.* they extend over the whole molecule. For closed shell systems, the SCF functions can be written in determinant form (Slater determinant). An important theorem states that certain linear transformations among the orbitals of the determinant can be carried out without changing the value of any physical observable. For the sake of simplicity, let us consider a *four-electron system* whose orbitals are labelled a and b and the spin orbitals $a\alpha$, $a\beta$, $b\alpha$, $b\beta$

$$\Phi = \frac{1}{\sqrt{4!}} \begin{vmatrix} a\alpha(1) & a\beta(1) & b\alpha(1) & b\beta(1) \\ a\alpha(2) & a\beta(2) & b\alpha(2) & b\beta(2) \\ a\alpha(3) & a\beta(3) & b\alpha(3) & b\beta(3) \\ a\alpha(4) & a\beta(4) & b\alpha(4) & b\beta(4) \end{vmatrix} \tag{4.5}$$

---

[a] The Hückel method predicts that in the case of a *long polyene chain* the longitudinal polarizability of π electrons varies as the cube of the molecular length, while the σ polarizability should increase only linearly [18]. Refined molecular orbital calculations (taking into account bond alternation etc.) do not exhibit this abnormal asymptotic behaviour.

One can convince oneself by calculating this determinant that the following function

$$\Phi = \frac{1}{\sqrt{4!}} \begin{vmatrix} A\alpha(1) & A\beta(1) & B\alpha(1) & B\beta(1) \\ A\alpha(2) & A\beta(2) & B\alpha(2) & B\beta(2) \\ A\alpha(3) & A\beta(3) & B\alpha(3) & B\beta(3) \\ A\alpha(4) & A\beta(4) & B\alpha(4) & B\beta(4) \end{vmatrix} . \tag{4.6}$$

with

$$A = a \cos \theta + b \sin \theta$$
$$B = a \sin \theta - b \cos \theta \tag{4.7}$$

(θ arbitrary)

is *exactly identical with the first one.* A Slater determinant is thus said to be 'invariant with respect to a unitary transformation among the occupied orbitals'.

Such a transformation can be used for relocalizing a given set of delocalized molecular orbitals in conformity with the chemical formula. For instance, the occupied orbitals of methane can be transformed into orbitals very close to simple two-center MO's constructed from tetrahedral sp^3 hybrid orbitals and $1s$ hydrogen orbitals [24,25,26]. A unitary transformation can hardly modify the wave function, except for an immaterial phase factor; therefore, it leads to a description which is as valid as that in terms of the 'canonical' delocalized Hartree-Fock orbitals. Of course, the localization obtained in this way is not perfect, but it is usually much better than is often believed. In the case of methane, the best 'localized' orbitals are uniquely determined by symmetry [27]; for less symmetric molecules one needs a criterion for best localization [28,29], a problem on which we shall not insist here. A careful inspection reveals that there are *three classes of compounds:*

i) Typical *non-conjugated molecules*, like ethylene or many other compounds, which everyone would write intuitively with localized bonds.

ii) Molecules usually classified by the chemists as *conjugated* but for which only one canonical valence formula (with the maximum number of double bonds) can be written, like the linear polyenes.

iii) *Conjugated molecules with two or more equivalent canonical valence formulas,* like benzene. In addition to their poor localizability, it can be shown that transformation to best localized orbitals is not unique [30].

One may now wonder what are the conditions for the occurrence of localized and delocalized bonds and what are the practical consequences of this more or less large localizability. The condition looked for was given a long time ago by Hund [31] (see also [32]).

The molecular orbitals can be transformed to localized ones if the number of valence electrons involved in binding per atom is equal to the number of directly bound neighbors and the number of atomic orbitals available.

In the case of carbon, there are four atomic orbitals, $2s$, $2p_x$, $2p_y$, $2p_z$, and four valence electrons. Therefore, carbon will form localized bonds if it is bound to four neighbors, each doubly-bonded neighbor being counted twice. In benzene, each carbon is bound to three of its neighbors by σ bonds and to two of them by an additional π bond; the number of neighbors is five and a description in terms of localized bonds is not possible[b]. In butadiene, on the other hand, the bond distances alternate in such a way that any C atom has only one neighbor close enough for a π bond, but three neighbors close enough for a σ bond; the number of neighbors is four and the MO's can be really well localized [33].

If a transformation to localized orbitals is possible, then the properties of the molecule can be described in terms of localized bonds, with some small correction for localization defects and interaction between the bonds (see Sect. 6.1). If no such transformation is possible, a description in terms of localized bonds does not lead to agreement with experiment, and auxiliary concepts like resonance or mesomerism have to be introduced in order to reconcile theory and experiment, unless one uses from the outset a description in terms of many-center bonds. From a practical point of view, the question why the π bonds are delocalized in benzene and localized in butadiene is reduced to the question *why the bond lengths are equal in benzene*, but not so in butadiene. Much work has been done on this question, but the situation is still being debated (see *e.g.* [34]).

Hund's criterion for localized bonds is well illustrated by the example of the BeH_2 molecule. The Be atom contributes two AO's, namely $2s$ and $2p_\sigma$, and has two neighbors and two valence electrons; each hydrogen has one neighbor and one electron. Hence, the bonds of BeH_2 can be localized. If one or two electrons are removed, the condition for localizability is no longer fulfilled, and the molecule can only be described by many-center bonds. The localized description also breaks down if no $2p_\sigma$ orbital is available, *i.e.* if we go from BeH_2 to HeH_2.

In triangular H_3^+ each hydrogen atom furnishes one AO and 2/3 of an electron, the number of neighbors being two for any hydrogen; so, the simplest delocalized bond is a σ bond. Actually, σ bonds in polyatomic molecules are only localized if they involve *e.g.* sp^n hybrids $(n = 1,2,3)$, *i.e.* if the AO basis functions can be transformed [35,36,37] in such a way that any of the resulting hybrid AO's can overlap in only one direction.

---

[b] Equivalent orbitals can be constructed even for benzene, but their definition is not unique [30] because different sets of equivalent orbitals are possible.

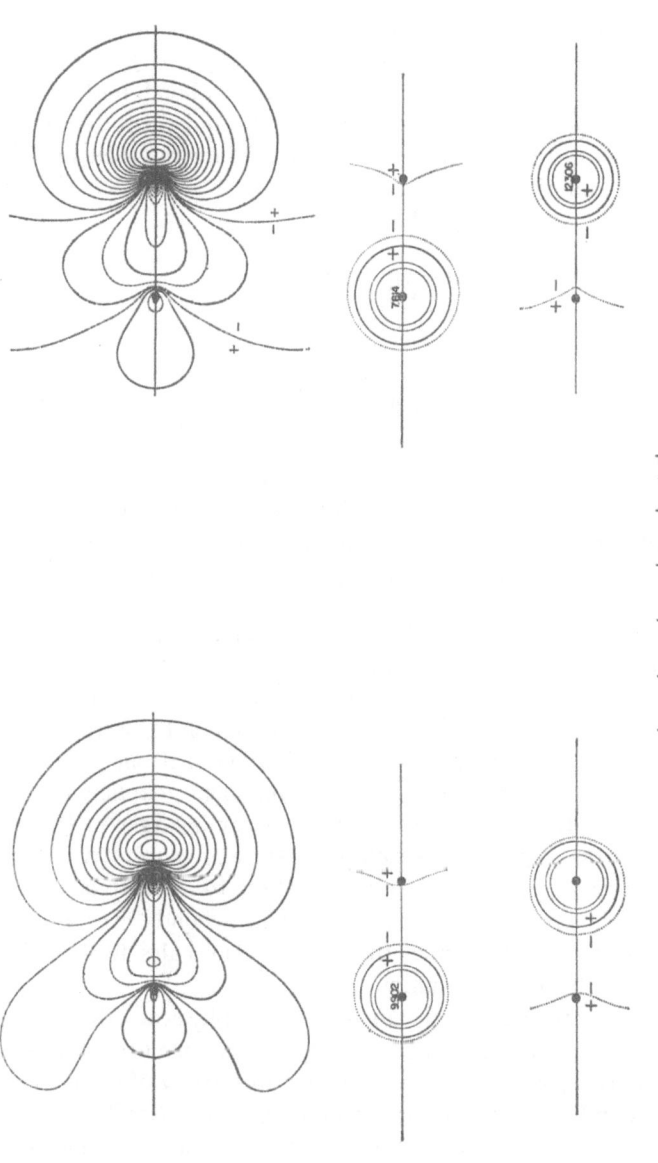

Fig. 4. Localized (equivalent) molecular orbitals in N_2 (left) and CO (right). (After K. Ruedenberg and L. S. Salmon, private communication). From top to bottom these MO's are: (1) one of three equivalent components of the triple bond, (2) the left and (3) the right lone pair, (4) the left and (5) the right inner shell orbitals

53

Pure s orbitals as well as π atomic orbitals can easily form many-center bonds, because they have the same binding power in different directions.

Having pointed out that a single Slater determinant is invariant with respect to a unitary transformation among the occupied orbitals, we can come to the idea of bent (or τ) bonds (see e.g. [38]). The double bond in ethylene is normally described by a σ and a π bond, both of them localized between the C atoms. Now, a completely equivalent description can be used where the bonding σ and π orbitals are replaced by their normalized sum and difference

$$\tau_1 = \frac{1}{\sqrt{2}} \, (\sigma + \pi)$$

$$(4.8)$$

$$\tau_2 = \frac{1}{\sqrt{2}} \, (\sigma - \pi)$$

These new orbitals look like *bent bonds* and are rather close to Van't Hoff's [39] original description of a double bond. The density diagrams of the various types of orbitals in diatomic molecules can be seen in 1 and 4, namely delocalized (canonical) MO's on Fig. 1 and localized (equivalent) MO's on Fig. 4.

Since in the framework of the independent particle model the double bond may equally well be described as a $\sigma-\pi$ or $\tau_1-\tau_2$ bond, the choice is mainly a matter of personal taste. However, there are two arguments in favour of the $\sigma-\pi$ picture. One argument is based on theoretical treatments going beyond the independent-particle model: if one leaves the IPM, the two descriptions are no longer equivalent; explicit calculations [40] show that the intrapair correlation energy is larger in the $\sigma-\pi$ picture, so that the latter is quantum-mechanically more satisfactory. The second argument is that a transformation to τ orbitals is only straightforward in systems with localized σ and π bonds, e.g. in molecules with isolated double bonds. In aromatic molecules like benzene the definition of τ bonds is not unique.

There is one interesting but much debated physical criterion for delocalized bonds in organic rings, namely the occurrence of the so-called *'ring currents'*, said to explain the high anisotropy of diamagnetic susceptibilities and some particular shielding effects in NMR spectra of aromatic molecules. The idea of the ring currents was developed by London [41] in an approximate treatment of the diamagnetic susceptibility of aromatic hydrocarbons. There has been some argument recently as to whether the ring currents are genuine physical effects or just an artefact of London's calculation [42,43,44].

4.6. Differences in Reactivity

It is well known that saturated molecules are rather unreactive and the reactivity in unsaturated and conjugated systems can be interpreted in terms of π electrons (see e.g. [45,46]). Nevertheless, the higher reactivity of π electron systems is not so much due to the different properties of σ and π electrons as to the fact that an unsaturated C atom has only three neighbors and can easily form a bond with a fourth atom in a transition state. In reactions characteristic of π systems, the σ core seems to play no role during the whole process, except for reactions involving a change of conformation. Quite often, to form a reaction intermediate from an unsaturated or conjugated molecule, it is not even necessary to break a π bond, so that the reaction intermediates as postulated, for instance, in aromatic substitution reactions have relatively low energy. The 'localization' energies [47] for the different possible intermediates allow us to discriminate between their energies and to find out which one will actually be preferred. Chemical reactions are, indeed, very complicated processes and it is fortunate that one can make a rather good classification in terms of quantities referring to the π electron system only.

To close this chapter on the differences between σ and π electrons, we emphasize that it is very useful to make the distinction between the two types of electrons, keeping in mind that 'electron' stands for 'occupied orbital'. From the point of view of quantum chemistry, this distinction is quite straightforward, but experiment does not provide really unambiguous ways of distinguishing between σ and π electrons and separating contributions due to the two sets. As a matter of fact, the π electrons seem to have the most important role for certain properties, which will be discussed in detail in the next chapter. However, the σ bonds in molecules with and without π electrons are not necessarily the same. In saturated hydrocarbons the σ bonds can be regarded as being formed by sp^3 hybrids, in unsaturated hydrocarbons by sp^2 hybrids. Some differences between the two classes of compounds are just due to differences in σ bonding (see e.g. [48]).

4.7. References

[1] Coulson, C. A., March, N. H., Altmann, S. L.: Proc. Nat. Acad. Sci. 38, 372 (1952).
[2] Newton, M. D., Boer, F. P., Lipscomb, W. N.: J. Am. Chem. Soc. 88, 2367 (1966).
[3] Slater, J. C.: Phys. Rev. 36, 57 (1930).
[4] Griffith, M. G., Goodman, L.: J. Chem. Phys. 47, 4494 (1967).
[5] Switkes, E., Stevens, R. M., Lipscomb, W. N.: J. Chem. Phys. 51, 5229 (1969).
[6] Van Vleck, J. H.: J. Chem. Phys. 2, 20 (1934).

[7] Mulliken, R. S.: J. Chem. Phys. *2*, 782 (1934).
[8] Pilcher, G., Skinner, H. A.: J. Inorg. Nucl. Chem. *24*, 937 (1962).
[9] Hinze, J., Jaffé, H. H.: J. Am. Chem. Soc. *84*, 540 (1962).
[10] Denbigh, K. G.: Trans. Faraday Soc. *36*, 936 (1940).
[11] Kołos, W., Wolniewicz, L.: J. Chem. Phys. *46*, 1426 (1967).
[12] Wilkins, R. L., Taylor, H. S.: J. Chem. Phys. *48*, 4934 (1968).
[13] Stevens, R. M., Lipscomb, W. N.: J. Chem. Phys. *41*, 184 (1964).
[14] O'Hare, J. M., Hurst, R. P.: J. Chem. Phys. *46*, 2356 (1967).
[15] McLean, A. D., Yoshimine, M.: J. Chem. Phys. *46*, 3682 (1967).
[16] Lefebvre, R. J. W.: Advan. Phys. Org. Chem. *3*, 1 (1965).
[17] London, F.: E. Z. Physik *63*, 245 (1930).
[18] Davies, P. L.: Trans. Faraday Soc. *48*, 789 (1952).
[19] Cohan, N. V., Coulson, C. A., Jamieson, J. B.: Trans. Faraday Soc. *53*, 582 (1957).
[20] Coulson, C. A., Longuet-Higgins, H. C.: Proc. Roy. Soc. (London) *A 191*, 39 (1947).
[21] Amos, A. T., Hall, G. G.: Theoret. Chim. Acta *6*, 159 (1966).
[22] Pople, J. A.: J. Chem. Soc. *37*, 53 (1962).
[23] Amos, A. T., Musher, J. I.: J. Chem. Phys. *49*, 2158 (1968).
[24] Coulson, C. A.: Trans. Faraday Soc. *38*, 433 (1962).
[25] Lennard-Jones, J. E.: Proc. Roy. Soc. (London) *A 202*, 1, 14 (1949).
[26] Lennard-Jones, J. E., Pople, J. A.: Proc. Roy. Soc. (London) *A 202*, 166 (1950).
[27] Hall, G. G.: Proc. Roy. Soc. (London) *A 202*, 336 (1950).
[28] Edmiston C., Ruedenberg, K.: Rev. Mod. Phys. *35*, 457 (1963).
[29] Foster, J. M., Boys, S. F.: Rev. Mod. Phys. *32*, 300 (1960).
[30] Edmiston, C., Ruedenberg, K.: in Quantum Theory of Atoms, Molecules and the Solid State, p. 263. New York: Academic Press 1966.
[31] Hund, F.: Z. Physik *74*, 1 (1931).
[32] Kimball, G. E.: J. Chem. Phys. *8*, 188 (1940).
[33] Staemmler, V., Kutzelnigg, W.: Theoret. Chim. Acta *9*, 67 (1967).
[34] Binsch, G., Heilbronner, E., Murrell, J. N.: Mol. Phys. *11*, 305 (1966).
[35] Del Re, G.: Theoret. Chim. Acta *1*, 188 (1963).
[36] Del Re, G., Esposito, U., Carpentieri, M.: Theoret. Chim. Acta *6*, 36 (1966).
[37] McWeeny, R., Del Re, G.: Theoret. Chim. Acta *10*, 13 (1968).
[38] Roberts, J. D., Caserio, M. C.: Basic Principles of Organic Chemistry. New York: Benjamin 1965.
[39] Van't Hoff, J. H.: Bull. Soc. Chim. *23*, 295 (1875).
[40] Klessinger, M.: J. Chem. Phys. *46*, 3261 (1967).
[41] London, F.: J. Phys. Rad. *8*, 397 (1937).
[42] Musher, J. I.: J. Chem. Phys. *43*, 4081 (1965); *46*, 1219 (1967).
[43] Gaidis, J. M., West, R.: J. Chem. Phys. *46*, 1218 (1967).
[44] Pople, J. A., Untch, K. G.: J. Am. Chem. Soc. *88*, 4811 (1966).
[45] Pullman, A., Pullman, B.: Les Théories Electroniques de la Chimie Organique. Paris: Masson 1952.
[46] Streitwieser, A.: Molecular Orbital Theory for Organic Chemists. New York: Wiley 1962.
[47] Wheland, G. W.: J. Am. Chem. Soc. *64*, 900 (1942).
[48] Dewar, M. J. S., Schmeising, H. N.: Tetrahedron *5*, 166 (1959).

5. π Electron Calculations and the Analysis of Experimental Data

5.1. Treatment of Unsaturated Molecules in Pure π-Electron Theories

If one considers only hydrocarbons, and more especially the so-called 'alternant' hydrocarbons, *i.e.* first of all the conjugated polyenes and the aromatic hydrocarbons of the benzene series, the greater part of their physical properties, ionization potentials, lower electronic transitions etc., can be interpreted qualitatively and often quantitatively in terms of the electronic structure of the π system alone. As the number of π electrons is small with respect to the total number of electrons of the molecule, a considerable simplification of the quantum-mechanical problem is obtained. However, it must be noted immediately that the assumptions of a complete $\sigma-\pi$ separation and of a rigid σ frame are not sufficient to eliminate the σ electrons completely from the theory, because the π electrons of an unsaturated molecule are not attracted by bare nuclei, but are subject to an effective potential containing Coulomb and exchange contributions from the σ electrons.

With the exception of semi-empirical methods based on the treatment of benzene given by Hückel in 1931 [1], all the π electron theories use more or less the procedure devised by Goeppert-Mayer and Sklar [2] (henceforth abbreviated GMS) to determine the potential in which the six π electrons of benzene move. As a rule, only the σ electrons belonging to the valence shell of the various atoms are taken into account; the electrons of the inner shells are assumed to shield the nuclei completely, *i.e.* to reduce, say, the nuclear charge $+6$ of carbon to $+4$. This simplification is not absolutely necessary (see *e.g.* [3]), but avoids some difficulties due to the orthogonality conditions between inner and outer orbitals. These conditions are automatically verified for the $2p\pi$ orbitals of the first-row atoms, which are orthogonal by symmetry to any $1s$, $2s$ and $2p\sigma$ orbitals; they are not in other cases, *e.g.* for the $3p\pi$ orbitals of sulphur and phosphorus, because in those cases there are π orbitals in the inner shells. Similar difficulties occur if one wants to calculate Rydberg transitions involving a σ-type excited orbital.

The basic GMS assumption lies in the form of the Hamiltonian H of the π electron system:

$$H = \sum_{\nu} h^{\text{core}}(\nu) + \sum_{\mu < \nu} \frac{e^2}{r_{\mu\nu}} \tag{5.1}$$

where the summations are taken only over the π electrons; the core Hamiltonian $h(\nu)$ is a one-electron operator containing, in addition to the customary kinetic and potential terms, an effective σ–π interaction potential assigned to the various atoms P of the molecule:

$$h^{\text{core}}(\nu) = T(\nu) + \sum_{P} U_P(\nu) \tag{5.2}$$

In order to give an explicit form to the potentials U_P, Goeppert-Mayer and Sklar assumed that the σ electron distribution around each atom is the same as in a molecule with infinitely large internuclear distances; the potential U_P is then given by the Hartree-Fock potential for the atom P in the appropriate valence state [4]; for instance, in the case of the carbon atom in the valence state (V_4, $s\,p_x\,p_y\,p_z$)

$$U_c(p_x) = -\frac{6}{r_e} + 2J_{1s} - K_{1s} + J_s + J_y + J_z - \frac{1}{2}(K_s + K_y + K_z)$$

$$\simeq -\frac{4}{r_e} + J_s + J_y + J_z - \frac{1}{2}(K_s + K_y + K_z) \tag{5.3}$$

where $J_i(\nu)$ and $K_i(\nu)$ are the Coulomb and exchange operator corresponding to each σ atomic orbital. Therefore, use of a GMS type potential amounts to neglecting the effect of σ bond formation on the electron structure of the π system, in particular, that of the intramolecular charge shifts in the σ system. It is possible to write the potentials U_P under much less severe conditions, provided that one knows the charges to be assigned to the σ electrons [5,6,7]. The results of complete calculations carried out recently by the MO method with Gaussian orbital basis sets suggest that the distribution of the gross atomic populations in aromatic molecules is not much different from the one which would be found by considering the ordinary valence states [8,9]. On the other hand, the picture of intramolecular charge transfers in heterocycles can be entirely modified; for instance, in aza-compounds the main part of calculated dipole moments seems to come from σ electrons [8,10] (see Sect. 6.4).

As has been mentioned in Chapter 3, the total wave function of the π-electron system is constructed from atomic orbitals that are antisymmetric with respect to the principal plane of the molecule. We shall confine ourselves to bases formed by the π valence orbitals of the unsaturated atoms of a molecule, e.g. the $2p_x$ orbitals of doubly linked carbon atoms. However, it is possible to generalize the treatment presented below to less familiar kinds of π orbitals, for instance the pseudo-atomic orbital $(1s_H - 1s_{H'})$, used account for the hyperconjugation of a saturated CH_2 group with a double bond [11]. Excited orbitals of the same symmetry as the π valence orbitals, like the $3p\pi$ [12,13] and $3d\pi$ [14] orbitals, could probably be included in a GMS treatment, but the theoretical meaning of these orbitals in the frame of a strict π-electron theory is not yet clear, even though they account for some experimental facts. For instance, Scheibe's rule [a] can be understood by incorporating $3p\pi$ orbitals in the π orbitals basis set [13,15,16,17], but this does not preclude other explanations [18].

The matrix elements of the total Hamiltonian H of the π-electron system include, first of all, the elements of the one-electron Hamiltonian h^{core} with respect to the valence orbitals χ_p of the π type centered on the different atoms P of the molecule. If there is only one orbital χ_p per unsaturated atom, one usually writes

$$[T(\nu) + U_P(\nu)]\,\chi_p(\nu) = W_p\chi_p(\nu) \tag{5.4}$$

grouping together the kinetic operator of electron ν and the potential operator associated to the atom P itself in the GMS approximation. Strictly speaking, the preceding equation means that the basis orbitals χ_p are eigenfunctions of a Hartree-Fock operator having the form $(T + U_P)$ and that W_p is the corresponding eigenvalue, $i.e.$ the energy the electron ν would have if it interacted only with the core of atom P in the valence state defined by the potential U_P. In practice, one represents the orbitals χ_p by simple algebraic expressions which are not really solutions of this equation; therefore, one should rather consider Eq. 4.1 as a symbolic relationship equivalent to the expression

$$\int\chi_p(\nu)[T(\nu) + U_p(\nu)]\,\chi_p(\nu)\,d\tau_\nu = W_p \tag{5.5}$$

and expand W_p in terms of atomic integrals over the basis functions χ_p.

A simpler extension of the GMS treatment was suggested by Moffitt [19] in connection with the calculation of the lower excited states of the oxygen molecule. This procedure only involves the valence state energies of the atom P and its positive and negative ions and π electronic integrals; it can easily be generalized to various kinds of unsaturated atoms [20,21, 22,23]. If the valence states of the ions P^+ and P^- obtained from atom

---

[a] Scheibe's rule states that the energy difference between the lowest excited state and the lowest ionized state in aromatic hydrocarbons is approximately equal to the same difference in the hydrogen atom.

P by extracting or fixing a π electron are constructed with the same orbital basis as those of P itself, the difference between the appropriate valence state energies can be expressed in terms of W_p; for an atom sharing two electrons with the π system (*e.g.* nitrogen in the $-NH_2$ group), or one electron (case of an ethylenic carbon), or having an empty $2p\pi$ orbital (boron in borazoles), W_p is respectively

$$W_p = E(P,V_i) - E(P^+, V_{i+1}) = -I_P - (p_x p_x; p_x p_x)$$

$$W_p = E(P,V_i) - E(P^+, V_{i-1}) = -I_P \qquad (5.6)$$

$$W_p = E(P^-, V_{i+1}) - E(P, V_i) = -A_P$$

where I_P and E_P are the ionization potential and the electron affinity of atom P for the $2p_x$ orbital in the valence state V_n, and

$$(p_x p_x; p_x p_x) = \int \chi_p(\mu)\, \chi_p(\mu)\, \frac{1}{r_{\mu\nu}}\, \chi_p(\nu)\, \chi_p(\nu)\, d\tau_\mu\, d\tau_\nu \qquad (5.7)$$

is the Coulomb repulsion integral for two π electrons described by an atomic orbital χ_p [22]. If all the atoms P are of the same kind (for instance, if all of them are carbon atoms), it is not necessary to give W_p a numerical value in order to calculate the wave function of the π-electron system, because W_P is the same in all the diagonal matrix elements and can be taken as the origin of the energy scale. In the case of substituted molecules, one only needs to know the values of the different W_P's in comparison to the W_P of the carbon atom. The required numerical values are usually taken from the experimental valence state energy tables [24, 25]. Such a procedure amounts to applying the approximation known as Koopmans' theorem (see Sect. 5.2) for the valence states of the atom P [21] An alternative procedure for evaluating the W_P's of atoms with lone pairs resorts to doubly excited valence states of P [26], but Koopmans' theorem cannot be extended to double excitations. Of course, the evaluation of the parameter W_p of boron (V_3, sp^2) should be less precise, because there is no counterpart to Koopmans' theorem for electro-affinities [21,27].

Note that the procedure described here can be extended without difficulty to atoms P which contribute several orbitals and several p electrons of different symmetries: carbon atoms in the sp hybridization state of acetylene and allene-type compounds [28], heteroatoms with one π electron and a lone pair, like oxygen in the carbonyl group or nitrogen in pyridine [20,21].

The non-diagonal elements of the Hamiltonian h^{core} contain integrals involving two functions χ_p and χ_q centered on two different atoms P and Q, and are usually written in the form

$$\int \chi_q(\nu)\,[T(\nu) + U_p(\nu)]\,\chi_p(\nu)\,d\tau_\nu = W_p S_{qp} \tag{5.8}$$

where the equality sign is obtained by assuming that χ_p is an eigenfunction of the Hartree-Fock operator for atom P. If χ_p is not an eigenfunction, the relation (5.8) can still be retained (with a somewhat different meaning for W_P) provided all the matrix elements in the brackets are proportional to the overlap integrals S_{pq}. However, this does not hold exactly for the basis functions that are used in practice: the most important deviation arises from the kinetic operator T, because the matrix elements T_{pq} between π orbitals are proportional to the square of the overlap S_{pq} [29]. An undesirable feature of the above relationship is that one obtains different values for the matrix elements with indices pq and qp, whenever the atoms P and Q are of a different nature, as in the case of carbon and oxygen in the carbonyl group: $W_p(C,V_4) \neq W_q(O,V_2)$. This difficulty does not arise if the Hamiltonian h^{core} is written in a symmetric form with respect to the potentials U_P and U_Q. *i.e.* if (supposing the overlap matrix S is real) one writes [20]:

$$\int \chi_p(\nu)\left[T(\nu) + \frac{U_P(\nu) + U_Q(\nu)}{2}\right]\chi_p(\nu)\,d\tau_\nu = \tfrac{1}{2} \times$$

$$\int \chi_p(\nu)\,[T(\nu) + U_P(\nu)]\,\chi_q(\nu) + \chi_q(\nu)\,[T(\nu) + U_Q(\nu)] \tag{5.9}$$

$$\chi_p(\nu)\,d\tau_\nu = \tfrac{1}{2}\,(W_P + W_Q)\,S_{pq}$$

A possible explanation of the inequality of the matrix elements with indices pq and qp can be given in terms of the σ electron distribution in the molecule: the polarity of the σ bond between the atoms P and Q affects the potentials U_P and U_Q in such a way that the symmetry of the core matrix is restored as a consequence of the σ–π interaction [30].

Given a specific atomic orbital basis, it is always possible to calculate all the matrix elements by integration, and to introduce the theoretical values found in that way as corrections to the terms $W_p S_{pq}$ of the ordinary GMS potential [31]. Unfortunately, these corrections are very sensitive to the choice of the basis, and it is difficult to give them a definite meaning.

Assuming that orbitals of the usual form (Slater orbitals etc...) are good approximations of the Hartree-Fock orbitals, the total matrix elements of the Hamiltonian h^{core} are

$$h_{pp} \simeq W_p + \int \chi_p \left(\sum_{R \neq P} U_R + \sum_N U_N \right) \chi_p \, d\tau = \alpha_p \qquad (5.10)$$

$$h_{pq} \simeq \tfrac{1}{2} (W_p + W_q) S_{pq} + \tfrac{1}{2} \int \chi_p (U_P + U_Q) \chi_q \, d\tau$$

$$+ \int \chi_p \left(\sum_{R \neq P, Q} U_R + \sum_N U_N \right) \chi_q \, d\tau = \beta_{pq} \qquad (5.11)$$

where the subscript R denotes an atom with π orbitals, and the subscript N any other atoms, for instance, the hydrogen atoms of an unsaturated carbon atom. It is convenient to replace the potential U_R produced by the nucleus and the σ electrons of the atom R by the potential U_R^0 of the atom R with all its electrons:

$$U_R^0 = U_R + n_R (J_x - \tfrac{1}{2} K_x) \qquad (5.12)$$

n_R being the number of π electrons on the $2 p_x$ orbital in the valence state V_i. Then, the following relation for the diagonal elements follows:

$$\alpha_p = W_p - \sum_{R \neq P} (U_R^0 ; pp) - n_R [(pp ; rr) - \tfrac{1}{2} (pr ; rp)] - \sum_N (U_N^0 ; pp) \qquad (5.13)$$

where

$$(U_R^0 ; pp) = - \int \chi_p U_R^0 \chi_p \, d\tau \qquad (U_N^0 ; pp) = - \int \chi_p U_N \chi_p \chi_p \, d\tau \qquad (5.14)$$

$$(pq ; rs) = \int \chi_p (\mu) \chi_q (\mu) \frac{1}{r_{\mu\nu}} \chi_r (\nu) \chi_s (\nu) \, d\tau_\mu \, d\tau_\nu \qquad (5.15)$$

and a similar expression for the non-diagonal elements β_{pq}. The quantities $(U_R^0 ; pp)$ are the familiar penetration integrals introduced by GMS in the π electron theories. They are of the short-range type, as opposed to the long-range forces of the Coulomb type $(pp ; rr)$ [29]; hence the approximate expression of semi-empirical methods:

$$\alpha_p \simeq W_p - \sum_{R \neq P} n_R (pp ; rr) \qquad (5.16)$$

In the particular case of hydrocarbons, it does not make much difference whether one neglects all the penetration integrals or retains only the penetration integrals $(U_C^0;pp)$ associated with the carbon atom neighboring the atom P and the integrals $(U_H^0;pp)$ associated with the hydrogen atoms bound to P. Since an sp^2 carbon atom has necessarily three neighbors, the latter approximation is practically equivalent to a shift of the origin of the energy scale, because $(U_C^0;pp)$ and $(U_H^0;pp)$ are not much different [32]. This circumstance is particularly favorable, because the numerical values of the penetration integrals differ whether the electric density of carbon is approximated by spherically charge distribution of the form

$$(2 s^2 + 2 p_x^2 + 2 p_y^2 + 2 p_z^2) \text{ [2]},$$

or if each of the integrals arising from U_C^0 is evaluated separately.

The calculation of the off-diagonal elements β_{pq} follows exactly the same line as that of the $\alpha_p's$, provided the Hamiltonian h^{core} is expanded symmetrically with respect to the potentials U_P and U_Q. If one assumes that all the one-electron and two-electron multi-center integrals contained in β_{pq} are expressed in terms of the corresponding Coulomb integrals appearing in α_p and α_q by means of the Mulliken approximation, one finds $\beta_{pq} = \frac{1}{2}(\alpha_p + \alpha_q) S_{pq}$. Actually, the preceding expression must be completed by a correction term ε_{pq}:

$$\beta_{pq} = \frac{1}{2}(\alpha_p + \alpha_q) S_{pq} + \varepsilon_{pq} \tag{5.17}$$

where ε_{pq} is the sum of the errors made by approximating each integral by Mullikens's formula. It may be noted that the important parameter of the π-electron theories is not so much the quantity β_{pq} itself as the correction term ε_{pq}, because one obtains no π binding at all by putting $\varepsilon_{pq} = 0$. In fact, if one replaces the basis orbitals χ_p by a set of orthogonalized Löwdin orbitals λ_p [33], one finds that the first term of the series expansion of this integral in powers of the overlap matrix [22,34] is

$$\beta_{pq}^{(\lambda)} = \int \lambda_p(\nu) \, h^{core}(\nu) \, \lambda_q(\nu) \, d\tau_\nu = \varepsilon_{pq} \tag{5.18}$$

Now, it is just this term which determines to a large extent the importance of the interaction of the π electrons between the atoms P and Q. The integral $\beta_{pq}^{(\lambda)}$ may be identified with the parameter β_{pq} of the semiempirical theories based on the zero-differential-overlap approximation [35,36]. In our opinion, there is no general calculation method leading to values for the β_{pq}'s which are in good numerical agreement with the β_{pq} parameters fitted on experimental data (electronic spectra, dipole moments etc.). Of course, opposite points of views on this controversial problem may be found in the literature [37].

We shall not discuss at length further simplifications, known as *next-neighbors interactions* (for core integrals), *zero-differential-overlap* (for two-electron repulsion integrals) etc., which were introduced into the formalism of the π molecular orbital theory after the basic work of Goeppert-Mayer and Sklar. Detailed reviews on these topics have been published [38,39,40,41,42]. Let us just show why zero-differential-overlap can be justified in terms of orthogonalized orbitals [43,22,44].

The Löwdin orthonormalized basis of atomic orbitals λ is obtained by the matrix transformation

$$\lambda = S^{-1/2} \chi \tag{5.19}$$

where χ is a basis of $2p\pi$ atomic orbitals localized on the various atoms of the molecule and S corresponding overlap matrix. The electron interaction integrals $(\lambda_p\lambda_q; \lambda_r\lambda_s)$ with respect to the λ basis can be expressed as matrix functions of the overlap and two-electron integrals in the χ basis. If the latter are calculated by Mulliken's approximation:

$$(\chi_p\chi_q; ..) = \tfrac{1}{2} S_{pq} [(\chi_p\chi_p; ..) + (\chi_q\chi_q; ..)] \tag{5.20}$$

except of course the Coulomb integrals $(\chi_p\chi_p; \chi_q\chi_q)$, one finds

$$(\lambda_p\lambda_q; ..) = 0 \text{ for } p \neq q \tag{5.21}$$

i.e. the properties postulated for the integrals in the zero-differential-overlap approximation. In fact, the preceding relationship is correct to the first order in the overlap integrals, and it can be shown to be rigorously correct, if the calculation is restricted to the term in S of the expansion in a series of the matrix $S^{-1/2}$, as a result of the use of Mulliken's formula for the non-Coulomb integrals. Now, Mulliken's approximation itself can be considered as the term in S for the expansion of such integrals [45]. So, the transformation of localized atomic orbitals into Löwdin delocalized orbitals gives to a first order in S a non-empirical formalism identical with the methods based on the zero-differential-overlap approximation suggested by Pariser, Parr and Pople.

Many interesting problems in physical organic chemistry have been clarified by numerical calculations based on next-neighbor interaction and zero-differential-overlap approximations, especially in the field of aromatic hydrocarbons. An important theorem of the Hückel theory for *hydrocarbons* holds also in theories of the Pariser-Parr-Pople type: alternant hydrocarbons do not exhibit π charge transfer from one unsaturated center to another [46,47]. This statement is valid not only for the ground state, but also for some excited states, in particular the lower singly excited state. Therefore the π distribution cannot induce any polarization in the σ system, and this means that the GMS potential is indeed a good approximation for such compounds. Actually, the special properties of conjugated hydrocarbons are determined by the topology of the molecule, and any method using the same diagonal matrix elements for atoms and arbitrary off-diagonal elements for chemically linked atoms is able to reproduce them [48].

A more physical interpretation of the potential to which the electrons of a conjugated molecule are subjected, was obtained by Del Re and Parr [49]. By transforming the expressions of the configuration interaction matrix elements over a molecular orbital basis, it is possible to show that the one-electron terms can be derived from an effective one-electron operator, where the potential is of the GMS type, but corresponds to a core including all the electrons but one, and equal fractions (per orbital) of the electrons are assigned to the various atoms. An important conclusion is that in large conjugated molecules with few *heteroatoms*, a π electron tends to see all the atoms as neutral, except for those heteroatoms that contribute two electrons to the π systems, which are seen as singly charged centers. Among other things, this suggests that the results concerning alternant hydrocarbons could be extended to a wider class of large conjugated molecules (with the exception of compounds with a highly polarized σ core), and thus may explain the success of the Hückel method even for heterocycles [50].

In the next two sections, we shall study the problem of the ionization potentials and electronic spectra of simple molecules, as examples of the virtues and shortcomings of π-electron theories for analyzing the characteristic properties of unsaturated molecules.

5.2. Ionization Potentials

Usually, theoretical studies on ionization processes of atoms and molecules are performed using the so-called approximation of *Koopmans' theorem*. This theorem says that the ionization potential P_i of an electron located on the ith level of a closed-shell state is equal to the opposite sign to the Hartree-Fock orbital energy e_i. One obtains this result by assuming that the single determinant wave function of the ion is constructed from the same molecular orbitals as the ground-state function, except for the spin orbital of the missing electron.

The ionization energy E_i, defined as the difference between the energy E^0 of the ground state and the energy E^+ of the positive ion, reduces then to the orbital energy e_i:

$$- P_i = E_i = E^0 - E^+ = e_i \qquad (5.22)$$

The evaluation of ionization potentials from orbital energies of the ground state is usually justified, not because ionization does not very much alter the form of orbitals, but because the values of e_i are in better agreement with experimental data than the difference of true SCF total energies, both for the positive ion and the neutral molecule. This statement is empirically verified for many atoms, but has to be theoretically explained by a fortunate cancellation of errors [51,52].

If one lets

$$P_i = E^+ - E^0 = (E_{SCF}^+ - E_{SCF}^0) + (E_c^+ - E_c^0) \qquad (5.23)$$

where E_c^+ and E_c^0 are the (negative) values of the correlation energies in the two systems, and

$$E_{SCF}^+ = E_{2n-1}^+ + \Delta E_i^+ \qquad (5.24)$$

where ΔE_i^+ is the additional (negative) error coming from the approximate calculation method used for the ion instead of a true SCF treatment, one can write

$$P_i = - e_i + \Delta E_i^+ + (E_c^+ - E_c^0) \qquad (5.25)$$

The correlation energy of a $2n$ electron system is in general larger in absolute value than that of the system with one less electron. Therefore, the quantity $(E_c^+ - E_c^0)$ is positive and tends to compensate the error ΔE_i^+. On the other hand, the same argument, applied to the calculation of electron affinities (the change in energy produced by the capture of an electron in an empty orbital φ_i), suggests that the errors $-\Delta E_i^-$ and $(E_c^0 - E_c^-)$ should cumulate rather than cancel.

The preceding discussion is relevant for vertical ionization potentials only, *i.e.* for ionization processes without a change in the shape of the molecule. The *effect of molecular distortions* due to ionization has to be estimated separately and added to ΔE_i^+ and $(E_c^+ - E_c^0)$. Recently, the validity of Koopmans' theorem has been questioned within the SCF approximation itself. It should be noted that the orbitals of a Slater determinant are defined except for a unitary transformation, and the canonical Hartree-Fock orbital energies give an upper bound to the difference between SCF total energies of both states only for the lowest energy level of each symmetry species [53]. Thus, the first π ionization potential of planar unsaturated molecules might safely be predicted. On the other hand, numerical SCF calculations have been performed for some ionized states of atoms [54,55] and molecules [56,57,58] and the cancellation of errors put forward to justify the use of Koopmans' theorem does not seem to be quite general. Furthermore, in the case of molecules, Table 6 shows that the basis set used for expanding the molecular orbitals plays an important role — indeed one, not yet well understood — as regards agreement with experiment.

Table 6. *First ionization potentials in simple molecules*

Basis set	$-e_1$ (in eV)				P_{exp}
	STO[1]	GTO[2]		OCO[3]	
		Limited	Extended		
Water	(11.79)	14.56	14.55	13.72	13.0
Ammonia	9.96	11.15	11.60	11.28	10.5
Methane	14.74	14.02	13.73	13.48	12.06
Ethane (stag.)	13.09	14.97	13.78	—	11.6
Ethylene	10.10	9.96	—	—	10.5
Acetylene	11.04	10.94	—	—	11.4

[1] Slater-type orbitals with $Z_H = 1.2$: Palke, W. E., Lipscomb, W. N.: J. Am. Chem. Soc. *88*, 2384 (1966). — $Z_H = 1$ for H_2O: Ellison, F. O., Shull, H.: J. Chem. Phys. *23*, 2348 (1955).
[2] Gaussian-type orbitals — H_2O, NH_3, CH_4: Ritchie, C. D., King, H. F.: J. Chem. Phys. *47*, 564 (1967). — C_2H_6: Clementi, E., Davis, D. R.: J. Chem. Phys. *45*, 2593 (1966). — C_2H_4; C_2H_2: Moskowitz, J.: J. Chem. Phys. *43*, 60 (1965).
[3] One-center orbitals: Moccia, R.: J. Chem. Phys. *40*, 2164, 2176, 2186 (1964).

In the lowest ionized state of *ethylene* and *acetylene*, a π electron belonging to the double or triple C—C bond is removed. It is interesting to compare the calculations reported in Table 6 where all electrons were included in the SCF treatment and the results of non-empirical calculations limited to the π electron system. Approximating the interaction of the π electrons with the σ core by a rigid GMS potential (see Sect. 5.1) and taking an effective nuclear charge equal to 3.18 for all the atomic integrals over 2π orbitals of carbon, the following ionization energies are found [59,60]:

$$\text{for the C=C bond: } e_i = W_{2p}^{tr} + 0.46 \text{ eV} = -10.70 \text{ eV}$$

$$\text{for the C}\equiv\text{C bond: } e_i = W_{2p}^{di} - 1.46 \text{ eV} = -12.65 \text{ eV}$$

where W_{2p}^{tr} and W_{2p}^{di} denote respectively the energy of a π electron belonging to a carbon atom in the sp^2 or sp hybridization:

$$W_{2p}^{tr} = -11.16 \text{ eV} \qquad W_{2p}^{di} = -11.19 \text{ eV [24]}$$

From a comparison with Table 6, one sees that the results of all-electron calculations are lower by about one eV than the ionization potentials predicted by pure π-electron theories. In addition, the meaning of the numerical value to be attributed to the additive constants W_{2p}, is somewhat obscure. This explains why these quantities may be regarded to a certain extent as adjustable molecular parameters, when ionization processes are the main object of a π-electron calculation [61,62].

Pure π-electron theories by definition neglect the change in the distribution of σ charges in the molecule. The possible role of such a change can be studied by simply performing SCF all-electron calculations for both states, because the average electronic repulsion given by the SCF method (see Chapt. 2.4.) is then different in the ground and ionized configurations. By solving separately the LCAO MO SCF equations of the neutral benzene and its $^2E_{1g}$ and $^2A_{2u}$ cations (*i.e.* benzene minus one electron in the first and second π energy levels), one finds that the polarity of the extracyclic σ bond is increased in the direction C⁻H⁺, so that the loss of a π electron belonging to the carbon atoms is balanced by a back donation of σ charges from hydrogens [56]. The same trend is shown when passing from pyridine to its $^2A_2$ cation (pyridine minus one π electron) [56]. Similarly, the charge transfer of the C–H bond increases from the cation CH_3^+ to the methyl radical and the anion CH_3^-, in proportion as the population of the $2p\pi$ orbital of carbon changes from zero to one and two [58]. On the other hand, the lowering of ionization potentials ΔE_i^+ obtained by minimizing the total energy of the ion is rather weak, as shown in Table 7.

For molecules with closed-shell ground states, the vertical ionization potentials $-e_i$ predicted by the LCAO MO method are generally too high relative to experiment. When the SCF energy of the ion is calculated, the correction ΔE_i^+ decreases the theoretical value of P_i, but the agreement with experiment is lost if the change in correlation energy $(E_c^+ - E_c^0)$ is taken into account in an approximate way (see Table 7).

Table 7. *Ionization potentials in benzene, pyridine and methyl*

Molecular ion	Ionization potentials in eV			
	$-e_i$ [1]	E_i^+	P_i [2]	P_{exp} [3]
Benzene cation $^2E_{1g}$	10.15	—0.41	11.74	9.25
Benzene cation $^2A_{2u}$	14.56	—0.44	16.12	12.1
Pyridine cation $^2A_2$	12.17	—0.71	13.46	9.24
Methyl cation $^1A_1'$	(10.18)	—1.32	8.84	9.84

[1] Gaussian-type orbitals — C_6H_6: Schulman, J. M., Moskowitz, I. W.: J. Chem. Phys. *47*, 3491 (1967) C_6H_5: Clementi, E.: J. Chem. Phys. *47*, 4485 (1967) CH_3 (e_i orbital energy of the open-shell Hamiltonian matrix): Millié, Ph., Berthier, G.: Int. I. Quant. Chem. *25*, 67 (1968).

[2] Correlation correction $(E_c—E_c^0)$ taken as equal to 2 eV for decoupling a pair of $2p$ electrons and neglected for methyl.

[3] C_6H_6: Turner, D. W.: Tetrahedron Letters *35*, 3419 (1967)
Lindholm, E., Jonsson, B. O.: Chem. Phys. Letters *1*, 501 (1967).
Momigny, J., Lorquet, J. C.: Chem. Phys. Letters *1*, 505 (1967).
C_6H_5 N: Momigny, J., Goffart, C., Natalis, P.: Bull. Soc. Chim. Belg. *77*, 533 (1968).
CH_3: Lossing, F. P., Ingold, K. U.: Henderson, I. H. S.: J. Chem. Phys. *22*, 621 (1954).

On the contrary, for *radicals*, the same type of correlation correction improves the agreement with experiment, because the difference between SCF total energies is too low. By optimizing the LCAO basis set and geometry used in the SCF calculation for the ion, there is hope that ionization potentials could be reproduced in a more satisfactory manner [57] [b].

5.3. Excitation Energies

As a general rule, saturated hydrocarbons do not absorb in the visible region or in the near and vacuum ultraviolet. The spectra of paraffins show absorption bands only in the far ultraviolet below 1500 Å. Two transitions are seen: the first, very weak, is apparently electronically forbidden: the second, more energetic and intense, is probably allowed

[b] Calculations of that sort, including a study of changes in correlation energies have been recently performed for water by W. Meyer and for methane by B. Levy, F. Janzat and J. Ridard (to be published). The results are very encouraging.

[63,64]. Therefore, the gap between the highest occupied energy level and the lowest empty level corresponding to the C—C and C—H bond system, in the frame of the simplest independent particle model, is at least nine eV. If the energy levels of σ bonds in saturated and unsaturated compounds are assumed to be of the same order of magnitude, the π electrons and the lone pairs of conjugated molecules must necessarily be assigned to transitions of lower energy observed from 1800—2000 Å to the near infrared.

In a one-electron scheme, there are three kinds of transitions involving π electrons:

the $\pi \rightarrow \pi^*$ transitions between the two π levels, one occupied and the other empty in the ground state of the molecule;
the $\pi \rightarrow \sigma^*$ and $\sigma \rightarrow \pi^*$ transitions between an occupied π level and an empty σ^* level, and *vice versa*;
the $n \rightarrow \pi^*$ transitions between a nonbonding n level occupied by the lone pair $2p\sigma$ of a heteroatom, as nitrogen of pyridine and oxygen of ketones, and an empty π^* level.

As the lone pairs are almost entirely localized on heteroatoms, the energy of the n level is almost the same as for an electron in the valence shell of the corresponding atom, so that the $n \rightarrow \pi^*$ and $n \rightarrow \sigma^*$ transitions have rather low energies and are often located in the same absorption region as pure π transitions. For instance, the absorption spectra of non-conjugated aldehydes and ketones include $n\text{-}\pi^*$ bands of very weak intensity around 3000 Å, more intense $n\text{-}\sigma^*$ bands around 2000 Å and strong $\pi\text{-}\pi^*$ bands below 1800 Å [65].

In compounds without lone-pair electrons, like pure hydrocarbons, the two or three highest occupied π levels and the two or three lowest empty π^* levels are respectively of higher and lower energy than the first σ and σ^* levels. Keeping in mind also that the empty levels form a more diffuse energy band than the occupied levels (a fact which is not predicted by the usual description using a basis set of atomic orbitals limited to valence orbitals of the various atoms), it is reasonable to assign the transitions in the visible and near ultraviolet to $\pi \rightarrow \pi^*$ excitations; the $\pi \rightarrow \sigma^*$ and $\sigma \rightarrow \pi^*$ transitions should be found only in the vacuum ultraviolet, the $\sigma \rightarrow \sigma^*$ transitions in the far ultraviolet. The sequence of occupied and empty levels for the valence electrons of the C≡C, C=C and C=O groups in the ground state is shown in Fig. 5, according to non-empirical SCF calculations carried out by a minimal Slater basis set for acetylene, ethylene [66] and formaldehyde [67].

The excitations considered so far are called valence shell transitions or *sub-Rydberg transitions*, as opposed to another type of excitations, found also in unsaturated compounds, namely the *Rydberg transitions* [68].

70

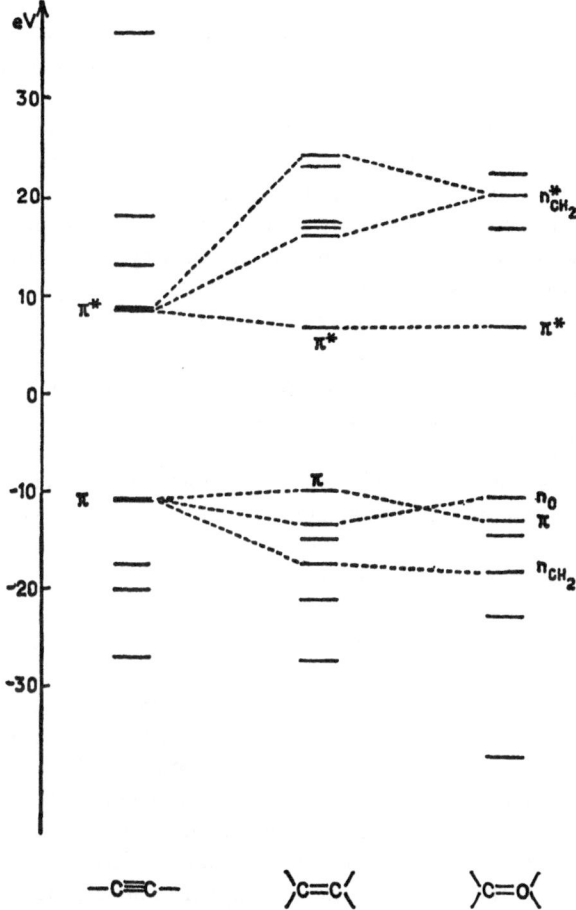

Fig. 5. Monoelectronic energy levels in acetylene, ethylene and formaldehyde

The ultraviolet spectrum of these compounds contains a series of bands, often very intense, which can be described by expressions of the form

$$h\nu = a - \frac{b}{(n-\alpha)^2} \tag{5.26}$$

where n is an integer and a, b, α are constants, as in the case of the atomic Rydberg series. The first terms of these series can reach the near ultraviolet; the next terms are in the far ultraviolet and converge towards a limit, which can be identified with a molecular ionization potential. The Rydberg series observed in unsaturated compounds are assigned to transitions between one of the last π levels of the ground state and a higher level where the excited electron takes a large probability density only very far from the molecular core. Roughly, the excited electron can be

71

described by a hydrogen-like wave function of high quantum number n ($n=3, 4, 5$, etc.) perturbed by the positive charge distribution of the residual molecular ion [69].

The Rydberg transitions are very close in energy to the $\pi \to \pi^*$ transitions lying in the vacuum ultraviolet; they must be distinguished from the latter by a careful analysis. Thus, the spectrum of ethylene below 1750 Å contains sharp Rydberg bands mixed with a continuum going up to 1620 Å; this continuum is the end of a system of diffuse bands going down to 2100 Å in gas phase and 2600 Å in liquid phase and is due to a $\pi \to \pi^*$ excitation with change in geometry [70]. From the theoretical point of view, there is not always a clear-cut distinction between valence shell transitions and Rydberg transitions, for instance, when the states in question belong to the same irreducible representations of the symmetry group of the molecule. This is the case for $n \to \sigma^*$ transitions of ketones, which can also be interpreted as $n \to 3s$ Rydberg transitions [71].

Another example is the puzzling problem of the 'mystery bands' in mono-olefins: the absorption spectrum of ethylenic compounds shows on the left side of the $\pi \to \pi^*$ band system weak bands or even simple shoulders of a different nature. These bands have been assigned to a transition involving π and σ levels simultaneously, and for some time a conflict raged amongst the supporters of a $\sigma \to \pi^*$ transition analogous to the well-known $n \to \pi^*$ band system of carbonyl compounds [72], those of a $\pi \to \sigma^*$ transition [73] and those emphasizing the absence of any mystery band in ethylene itself [74,75]. The dispute has been provisionally settled by assigning the mystery band of mono-olefins to the jump of a π electron towards a σ^* level strongly mixed with the term $3s$ of a Rydberg series [76,77]. Furthermore, a third low-lying excited state, located at 7.45 eV above the ground state in ethylene and electric-quadrupole-allowed [77,78], has been identied as a π transition towards a σ $3p$ Rydberg level. It can be added that, in the united atom model, all the transitions are interpreted as Rydberg transitions: for instance, the $\pi \to \pi^*$ excitation of the C=C double bond electrons becomes a $2p_x - 3d_{yz}$ transition of the united atom. Nevertheless, the usual classification retains a great utility, owing to the fact that the excitations usually interpreted as valence shell transitions are related more closely to the electronic structure of the molecule under consideration and cover a much larger spectral region, from the far ultraviolet to the near infrared.

In the preceding descriptions, the electrons have been assigned to individual energy levels. In fact, it is only possible to observe changes concerning the state of the whole molecule, and the energy of a transition is the difference between the final energy and the initial energy of the whole electron system. Even if one assumes that the excitation does not modify the position of the one-electron energy levels attributed to the ground state by the IPM model, the energy of a transition is not simply given by the energy difference between the final and initial levels, because the Coulomb repulsion and the spin coupling of the electrons involved in the transition are not the same in the two states. In molecules the most common case is a closed-shell ground state, and an excited state with two unpaired electrons, one on a molecular orbital φ_i, which was originally doubly occupied, the other on a molecular orbital φ_j, which was empty. The energy of the excited state depends on the total spin of the two unpaired electrons, whence the possibility of singlet-singlet transitions without any change in spin and of singlet-triplet transitions with change in spin. If the occupied orbitals φ_i and the empty orbitals φ_j are chosen from amongst the solutions of the Hartree-Fock equations of the ground state, the excitation energy $i \rightarrow j$ can be written [79]:

$$^{1,3}E - E_0 = (e_j - e_i) - (J_{ij} - K_{ij}) \pm K_{ij} \tag{5.27}$$

with the plus sign for the singlet and the minus sign for the triplet; the quantities e_i and e_j are the energies of the orbitals φ_i and φ_j involved in the transition, J_{ij} is the Coulomb repulsion integral between the charge distributions $\varphi_i^* \varphi_i$ and $\varphi_j^* \varphi_j$, and K_{ij} is the corresponding exchange integral. The quantity $2K_{ij}$ represents the singlet-triplet separation, the triplet state being below the singlet excited state, in agreement with Hund's rule $(0 \leqslant K_{ij} \leqslant J_{ij})$.

Although the singlet-triplet transition is spin-forbidden, the *singlet-triplet separation* has a considerable theoretical and practical interest: the value of K_{ij} differs very much according to the probability of finding the unpaired electrons in the same region of the three-dimensional space. In general, the transitions involving molecular orbitals of the same type ($\pi \rightarrow \pi^*$ and $\sigma \rightarrow \sigma^*$) are characterized by relatively large singlet-triplet separations. The same rule holds for the intensities, unless the transition is forbidden for reasons due to the molecular geometry (as is the case with some transitions for compounds of high symmetry such as acetylene and benzene) or to the spin (case of singlet-triplet transitions). For instance, the excitation of the π electrons of the C=C double bond in ethylene gives rise to a triplet and a singlet π-π^* located respectively at 7.6 eV and 4.6 eV with respect to the ground state [80]; the intensity of the singlet-singlet transition, measured by its oscillator strength [c], is equal to 0.3 [81].

---

[c] The oscillator strength f is defined as the ratio of the probability of a given transition to that of a harmonic oscillator able to absorb the same electromagnetic energy between its ground state and its first excited state. Quantum-mechanically, the transition probability is proportional to the square of the so-called 'transition moment'.

The transitions $\pi \to \sigma^*$ or $\sigma \to \pi^*$ have much lower singlet-triplet separations and intensities. The extreme limit is shown by the orbitals of a lone pair, which are σ orbitals almost completely localized on an atom. In the case of the carbonyl group, the orbital usually assigned to the lone pair n of oxygen is a $2p_y$ orbital perpendicular to the orbitals of the double bond C=O; the triplet and the singlet states n-π^* are extremly close to each other, and the n-π^* transition is locally forbidden, therefore very weak ($f \sim 0.01$ for singlet-singlet transitions)[20].

As shown in Table 8, it is not possible to account for the transitions involving the π electrons of the simplest organic molecules by non-empirical calculations based on an independent particle model. Both the singlet-triplet separation and the oscillator strength of the $\pi \to \pi^*$ transition are overestimated.

Table 8. *Transition energies of acetylene, ethylene and formaldehyde*

Excited states in eV		π electrons[1]		All electrons[2]		Exp.[3]
		SCF	Full CI	SCF	CI	
Acetylene						
$\pi \to \pi^*$	$^3\Sigma_u^+$	3.73	5.06			
	$^3\Delta_u$	4.77	6.46			
	$^3\Sigma_u^-$	5.81	7.78			
	$^1\Sigma_u^-$	5.81	7.78			5.2
	$^1\Delta_u$	6.43	8.60			
	$^1\Sigma_u^+$	17.80	15.81			7.9
Ethylene						
$\pi \to \pi^*$	$^3B_{3u}$	1.8	3.1	3.36	3.19	4.6
	$^1B_{3u}$	10.2	11.5	11.98	10.17	7.6
Formaldehyde						
$n \to \pi^*$	$^3A_2$	4.83	4.84	2.88	2.33	3.1
	$^1A_2$	6 01	5.26	4.03	3.60	3.5
$\pi \to \pi^*$	$^3A_1$	3.90	5.24	3.99	3.88	
	$^1A_1$	14.60	15.44	14.89	12.03	7.9

[1] Non-empirical calculations with $2p\pi$ Slater orbitals.
  C_2H_2: Serre, J.: J. Chimie phys. *50*, 447 (1953) and Thesis, Paris (1955).
  C_2H_4: Parr, R. G., Crawford, B. L.: J. Chem. Phys. *16*, 526 (1948).
  H_2CO: Sender, M., Berthier, G.: J. Chimie phys. *56*, 946 (1959).
[2] Complete calculations with a Slater minimal basis set CI limited to singly excited configurations (Tamm-Dankoff Approximation).
  C_2H_4: Dunning, T. H., McKoy, V.: J. Chem. Phys. *47*, 1735 (1967).
  H_2CO: Dunning, T. H., McKoy, V.: J. Chem. Phys. *48*, 5263 (1968).
[3] Peak of absorption maximum.

No improvement is obtained by replacing pure π electron calculations by an all-electron SCF treatment; thus, the assumption of a GMS potential can hardly be responsible for the failure of the theory. Furthermore, the theoretical predictions are not significantly altered by a configuration interaction limited to the $2p\pi$ orbital basis [59,83,84,85]; only by combining the $\sigma \rightarrow \sigma^*$ excitations and $\pi \rightarrow \pi^*$ excitations with the same symmetry properties can one succeed in reducing the $\pi \rightarrow \pi^*$ transition energy [86,87,88]. Therefore, a satisfactory *ab initio* theory of spectra will probably need to go beyond the $\sigma - \pi$ separation (see Sect. 6.4).

There are other possible reasons for the discordance between simple theories and experimental facts. First of all, much caution is needed in comparing theoretical and experimental excitation energies: the values calculated are usually vertical transition energies in the sense of the Franck-Condon principle, *i.e.* transitions without a change in geometry between the ground and excited states. They should not be compared with the energies of the experimental zero-zero transitions, but rather with the absorption maxima λ_{max}. It may happen that the *equilibrium geometry* of the molecule is completely different in the ground and excited states; this is the case for the first excited singlet of acetylene (which is bent [89]), ethylene (which is twisted and perhaps pyramidal [70]) and formaldehyde (which is pyramidal [90]). Changes in geometry are not always so drastic, especially in cyclic molecules like benzene (which is only slightly distorted in its first singlet excited state [91,92]); nevertheless, they cannot be accounted for by simply modifying the lengths of the unsaturated bonds in a pure π calculation. To take into account, say, changes in shape of acetylene [93] or ethylene [94] in their excited states, it is necessary to include at least the σ electrons of single bonds adjacent to the π bond system.

Various procedures have been suggested for improving the calculation of $\pi \rightarrow \pi^*$ transitions without modifying the general interpretation of spectra. One of them consists in taking the effective charges of the orbitals as additional variational parameters, whose values could be different for the $2p\pi$ and core orbitals. Several procedures based on that idea have been developed: variation of the $2p\pi$ orbital exponents according to the nature of the spectroscopic state under study [95], the ionic or covalent character of the valence bond structures of the molecule [96]; the bonding or antibonding character of the molecular orbitals occupied by the two unpaired electrons [97]. The last treatment, first suggested for the $\sigma \rightarrow \sigma^*$ transition of hydrogen molecule [98], has interesting connections with the form of orbitals in excited states. It is found that the effective charges of the $2p\pi$ carbon orbitals in ethylene are much smaller for the excited level π^* than for the ground level π. This means that the antibonding molecular orbital of a double bond is more spread out in space than the bonding one, a fact which is not recognized by any LCAO MO theory using minimal orbital basis sets, *e.g.* one $2p\pi$ orbital per carbon atom and π electron. As a matter of fact, the calculation of $\pi \rightarrow \pi^*$ transitions can be improved in a more conventional way by taking expansion bases with a large number of orbitals of π symmetry [99,100].

At indefinitely large internuclear distances, any molecular wave function coming from a full configuration interaction treatment can be expressed in terms of atomic valence state wave functions. According to Moffitt [101], the formation of chemical bonds can be regarded as a perturbation acting on isolated atoms, and the failure of π theories is apparently due to the fact that the energy spectrum of the dissociation products of the molecule is poorly represented by the usual methods of quantum chemistry. The excitations predicted for the molecule at equilibrium distances are much more satisfactory if the Hamiltonian matrix elements at infinity are replaced by spectroscopic valence state energies [102,103]. Later considerations [104,105,106] showed that Moffitt's method of 'atoms in molecules' could not be developed in a rigorous way, except for systems with very few electrons. When the interpretation of experimental data is the main object of theoretical calculations, it is more economical and fruitful to incorporate the data of atomic spectroscopy into a purely empirical scheme. This approach is the basis of the well-known method of Pariser and Parr [35] and Pople [36]: by fitting a number of basic core integrals β_{pq} and bielectronic repulsion integrals $(pp; pp)$ and $(pp; qq)$, it is possible to reproduce molecular spectra much better than by any theoretical method. Thus, the sequence of transitions in a given compound, say the benzene molecule (see Table 9), or the variation of a transition in a set of related molecules (see [107,108]) can be succesfully predicted. Such calculations have certainly a great heuristic value; however, they include the effects of the atomic basis set, the average Coulomb σ-π interaction and the σ-π and π-π electron correlation in a way not susceptible of theoretical analysis.

To summarize, ionization and excitation energies support the familiar picture of unsaturated molecules in terms of π electrons. Nevertheless, we wish to stress the point that agreement or disagreement with experiment by no means proves or disproves an approximate theory. There is often an alternative explanation for the characteristics of unsaturated compounds: in Chapt. 4.5, we haved noted that the properties related to the delocalization of π electrons (ring currents etc. . .) could be interpreted in a different way, even for benzene. Curiously enough, the electronic transition of benzene can be reproduced by a GMS treatment involving the σ electrons of C—C and C—H bonds instead of the π electrons of the ring [109].

Table 9. *Lower transition energies of benzene*

Excited states in eV	π electrons [1]				All electrons [2]		Exp. [3]
	SCF a)	CI a)	Moffitt b)	PPP c)	SCF	CI	
Triplets							
$^3B_{1u}$	3.07	2.56	5.2	3.59	4.54	3.98	3.66
$^3E_{1u}$	4.32	3.82	5.6	4.15	5.73	5.39	4.69
$^3B_{2u}$	5.56	7.95	6.1	4.71	6.92	8.61	5.76
Singlets							
$^1B_{2u}$	5.80	3.58	5.8	4.71	7.15	5.26	4.89
$^1B_{1u}$	7.10	8.77	4.8	5.96	8.38	9.48	6.14
$^1E_{1u}$	9.76	9.69	8.3	6.55	10.93	10.61	6.75

[1] $2p\pi$ Slater orbitals — a) Non-empirical claculations: Moskowitz, J. W., Barnett, M. P.: J. Chem. Phys. *39*, 1557 (1963). b) Atoms in Molecules: Moffitt, W., Scanlan, J.: Proc. Roy. Soc. (London): *A 220*, 530 (1953) c) Semi-empirical claculations with CI limited to singly excited configurations: Pariser, R.: J. Chem. Phys. *24*, 250 (1956).

[2] Gaussian orbitals: full CI limited to the π electron system. Buenker, R. J., Whitten, J. L., Petke, J. D.: J. Chem. Phys. *49*, 2261 (1968).

[3] Kearns, D. R.: J. Chem. Phys. *36*, 1608 (1962).

5.4. References

[1] Hückel, E.: Z. Physik *70*, 204 (1931).
[2] Goeppert-Mayer, M., Sklar, A. L.: J. Chem. Phys. *6*, 645 (1938).
[3] Suard-Sender, M.: J. Chim. Phys. *64*, 79 (1965).
[4] Moffitt, W.: Proc. Roy. Soc. (London) *A 202*, 534 (1950).
[5] Julg, A.: J. Chim. Phys. *55*, 413 (1958).
[6] Jungen, M., Labhart, H.: Theoret. Chim. Acta *9*, 345, 366 (1968).
[7] Denis, A., Gilbert, M.: Theoret. Chim. Acta *11*, 31 (1968).
[8] Clementi, E.: J. Chem. Phys. *46*, 4725, 4731, 4737 (1967).
[9] Praud, L., Millié, Ph., Berthier, G.: Theoret. Chim. Acta *11*, 169 (1968).
[10] Berthier, G.: Chimia *22*, 385 (1968).
[11] Levy, B., Berthier, G.: J. Chim. Phys. *63*, 1375 (1966).
[12] Jacobs, J.: Proc. Phys. Soc. (London) *68*, 72 (1955).
[13] Hartmann, H.: Z. Naturforsch. *15*a, 993 (1960).
[14] Sowers, O., Kauzmann, W.: J. Chem. Phys. *38*, 813 (1963).
[15] Preuss, H.: Z. Naturforsch. *16a*, 800 (1961).
[16] Ruch, E.: Z. Naturforsch. *16a*, 808 (1961).
[17] Bingel, W., Preuss, H., Schmidtke, H. H.: Z. Naturforsch. *16*a, 1328.
[18] Kollaard, U. K., Colpa, J. P.: Mol. Phys. *8*, 295 (1964).
[19] Moffitt, W.: Proc. Roy. Soc. *A 210*, 224 (1951).
[20] Sender, M., Berthier, G.: J. Chim. Phys. *55*, 384 (1958).
[21] Anno, T.: J. Chem. Phys. *29*, 1161 (1958).

22) Berthier, G., Suard, M., Baudet, J.: Tetrahedron *19*, Suppl. 2, 1 (1963).
23) Leroy, G.: Bull. Soc. Chim. Belg. *73*, 166 (1964).
24) Hinze, H., Jaffe, H. H.: J. Am. Chem. Soc. *84*, 540 (1962).
25) Pilcher, G., Skinner, H.: J. Inorg. Nucl. Chem. *24*, 937 (1962).
26) Paoloni, L.: Nuovo Cimento X, *4*, 410 (1956).
27) Lesk, A. M.: Phys. Rev. *171*, 7 (1968).
28) Serre, J.: J. Chim. Phys. *51*, 568 (1954); *53*, 284 (1956).
29) Ruedenberg, K.: J. Chem. Phys. *34*, 1861 (1961).
30) Julg, A., Bonnet, M.: Theoret. Chim. Acta, *1*, 1 (1962).
31) Stewart, T. L.: Proc. Phys. Soc. (London) *A 65*, 220 (1960).
32) Julg, A., Pullman, B.: J. Chim. Phys. *52*, 481 (1955).
33) Löwdin, P. O.: J. Chem. Phys. *18*, 365 (1950).
34) Peradejordi, F.: Compt. Rend. *243*, 276 (1956).
35) Pariser, R., Parr, R. G.: J. Chem. Phys. *21*, 466, 767 (1953).
36) Pople, J. A.: Trans. Faraday Soc. 49, 1375 (1953).
37) Ohno, K.: Theoret. Chim. Acta 2, 219 (1964).
38) Lykos, P. G.: Advan. Quant. Chem. *1*, 171 (1964).
39) I'Haya, Y.: Advan. Quan. Chem. *1*, 203 (1964).
40) Fischer-Hjalmars, I.: Advan. Quan. Chem. 2, 25 (1965).
41) Ohno, K.: Advan. Quan. Chem. *3*, 240 (1967).
42) Jug, K.: Theoret. Chim. Acta *14*, 91 (1969).
43) Fumi, F. G., Parr, R. G.: J. Chem. Phys. *21*, 1864 (1953).
44) Fischer-Hjalmars, I.: J. Chem. Phys. *42*, 1962 (1965).
45) Ruedenberg, K.: J. Chem. Phys. *19*, 1433 (1951).
46) Coulson, C. A., Rushbrooke, G. S.: Proc. Cambr. Phil. Soc. *36*, 193 (1940).
47) McLachlan, A. D.: Mol. Phys. 2, 361 (1959).
48) Ruedenberg, K.: J. Chem. Phys. *34*, 1884 (1961).
49) Del Re, G., Parr, R. G.: Rev. Mod. Phys. *35*, 604 (1963).
50) Carpentieri, M., Porro, L., Del Re, G.: Int. J. Quant. Chem. 2, 807 (1968).
51) Mulliken, R. S.: J. Chim. Phys. *46*, 497 (1949).
52) Moffitt, W.: Proc. Roy. Soc. (London) *A 202*, 534 (1950).
53) Newton, M. D.: J. Chem. Phys. *48*, 2825 (1968).
54) Sureau, A., Berthier, G.: J. Physique *24*, 672 (1963).
55) Bagus, P. S.: Phys. Rev. *139*, A 619 (1965).
56) Schulman, J. M., Moskowitz, J. W.: J. Chem. Phys. *47*, 3491 (1967).
57) Clementi, E.: J. Chem. Phys. *47*, 4485 (1967).
58) Millié, Ph., Berthier, G.: Int. J. Quant. Chem. 2S, 67 (1968).
59) Serre, J., Pullman, A.: J. Chim. Phys. *50*, 447 (1953).
60) Serre, J.: Thèse, Paris (1955).
61) Sidman, J. W.: J. Chem. Phys. *27*, 429 (1957).
62) Pullman, A., Rossi, M.: Biochim. Biophys. Acta *88*, 211 (1964).
63) Lombos, B. A., Sauvageau, P., Sandorfy, C.: Chem. Phys. Letters *1*, 42, (1967).
64) Raimonda, J. W., Simpson, W. T.: J. Chem. Phys. *47*, 430 (1967).
65) McMurry, H., Mulliken, R. S.: Proc. Natl. Acad. Sci. *26*, 312 (1940).
66) Palke, W. E., Lipscomb, W. N.: J. Am. Chem. Soc. *88*, 2384 (1966).
67) Newton, M. D., Palke, W. E.: J. Chem. Phys. *45*, 2329 (1966).
68) Mulliken, R. S., Rieke, C. A.: Rep. Prog. Phys. *8*, 231 (1941).
69) Liehr, A. D.: Z. Naturforsch. *11 a*, 752 (1956).
70) Wilkinson, P. G., Mulliken, R. S.: J. Chem. Phys. *23*, 1895 (1955).
71) Mulliken, R. S.: quoted by H. Tsubomura in Bull. Chem. Soc. Japan *37*, 417 (1964).
72) Berry, R. S.: J. Chem. Phys. *38*, 1934 (1963).

[73] Robin, M. B., Hart, R. R., Kuebler, N. A.: J. Chem. Phys. *44*, 1803 (1966).

[74] Evans, D. F.: J. Chem. Soc. 1735 (1960).

[75] Lubezky, A., Kopelman, R.: J. Chem. Phys. *45*, 2526 (1966).

[76] Robin, M. A., Basch, M., Kuebler, N. A., Kaplan, B. E., Meinwald, J.: J. Chem. Phys. *48*, 5037 (1968).

[77] Yaris, M., Moscowitz, A., Berry, R. S.: J. Chem. Phys. *49*, 3150 (1968).

[78] Ross, K. J., Lassettre, E. N.: J. Chem. Phys. *44*, 4633 (1966).

[79] Roothaan, C. C. J.: Rev. Mod. Phys. *23*, 69 (1951).

[80] Mulliken, R. S.: J. Chem. Phys. *33*, 1596 (1960).

[81] Zelikoff, M., Watanabe, K.: J. Opt. Soc. Am. *43*, 756 (1953).

[82] Berthier, G., Serre, J.: In: The Chemistry of the Carbonyl Group; edited by S. Patai, Interscience (1966).

[83] Parr, R. G., Crawford, B. L.: J. Chem. Phys. *16*, 526 (1948).

[84] Sender, M., Berthier, G.: J. Chim. Phys. *56*, 946 (1959).

[85] Kaldor, U., Shavitt, I.: J. Chem. Phys. *48*, 191 (1968).

[86] Dunning, T. H., McKoy, V.: J. Chem. Phys. *47*, 1735 (1967).

[87] — McKoy, V.: J. Chem. Phys. *48*, 5263 (1968).

[88] Denis, A., Malrieu, J. P.: J. Chem. Phys. *52*, 4762, 4769 (1970).

[89] King, G. W., Ingold, C. K.: J. Chem. Soc. 2740 (1953).

[90] Brand, J. C. D.: J. Chem. Soc. 852 (1956).

[91] Garforth, F. M., Ingold, C. V., Poole, H. G.: J. Chem. Soc 508 (1948).

[92] Craig, D. P.: J. Chem. Soc. 2146 (1960).

[93] Howard, H., King, G. W.: Can. J. Chem. *37*, 700 (1959).

[94] Burnelle, L., Litt, C.: Mol. Phys. *9*, 433 (1965).

[95] Murai, T.: Prog. Theoret. Phys. *7*, 345, (1952).

[96] Ohno, K., Itoh, T.: J. Chem. Phys. *23*, 1468 (1955).

[97] Huzinaga, S.: J. Chem. Phys. *36*, 453 (1962).

[98] Phillipson, P. E., Mulliken, R. S.: J. Chem. Phys. *28*, 1248 (1958).

[99] McKoy, V.: (unpublished results).

[100] Denis, A., Malrieu, J. P.: J. Chem. Phys. *52*, 6076 (1970).

[101] Moffitt, W.: Proc. Roy. Soc. (London) *A 218*, 486 (1953).

[102] Moffitt, W., Scalan, J.: Proc. Roy. Soc. (London) *A 218*, 464 (1953), *A 220*, 530, (1953).

[103] Serre, J.: Compt, Rend. *242*, 1469 (1956).

[104] Pauncz, R.: Acta Phys. Hung. *4*, 237 (1954).

[105] Hurley, A. C.: Proc. Phys. Soc. (London) *A 68*, 149 (1955).

[106] Arai, T.: J. Chem. Phys. *26*, 435, 451, (1957); *38*, 32 (1958).

[107] Koutecky, J., Paldus, J., Zahradnik, R.: J. Chem. Phys. *36*, 3129, (1962).

[108] Zahradnik, R.: Fortschr. Chem. Forschg. *10*, 1 (1968).

[109] Paoloni, L., De Giambiagi, M. S., Giambiagi, M.: Atti Soc. Nat. Mat. Modena *100*, 89 (1969).

6. Inclusion of σ Electrons in Molecular Calculations

6.1. σ Electron Theories

Until about 1960 theoreticians devoted much less attention to σ electron systems than to π electron systems. The reason was mainly that the number of σ electrons in any molecule is very large compared with the number of π electrons in conjugated systems. Actually, it is no more difficult to elaborate simplified calculation methods for σ electrons than for π electrons: indeed, such treatments had been suggested many years before (see [1] for a review of the early period and [2] for more recent developments).

The progress of digital computers in the last ten years has made it possible to calculate approximate wave functions for rather large electron systems; for instance, σ and π electrons of biological molecules, such as the fundamental bases of nucleic acids, have been treated not only by semi-empirical methods, but also by *ab initio* methods including all electrons[a] [3,4]. Nevertheless, one should not overestimate the accuracy of the calculations that have been carried out so far. As regards energy and related observables (binding and excitation energies, force constants, etc. . .), one possible, if not entirely satisfactory, classification of quantum-mechanical treatments may be given according to the order of magnitude of the error ΔE in the total energy:

i) spectroscopically accurate calculations: $\Delta E \sim 1\,\text{cm}^{-1}\,(3 \cdot 10^{-3}\,\text{kcal} \cdot \text{mol}^{-1})$

ii) chemically accurate calculations: $\Delta E \sim 1\,\text{kcal} \cdot \text{mol}^{-1}$

iii) moderately accurate calculations: $\Delta E \sim 1\,\text{eV}$ ($23\,\text{kcal} \cdot \text{mol}^{-1}$)

iv) crude calculations: $\Delta E \sim 100\,\text{kcal} \cdot \text{mol}^{-1}$.

---

[a] *Ab initio* calculations are quantum-mechanical treatments performed from first principles. In the case of molecules, the only input data are the number and nature of nuclei and the number of electrons, in other words the gross chemical formula. No other experimental data should be used, but the calculations are often limited to experimental geometries. It should also be stated that so-called *ab initio* calculations are usually approximate and often have a highly empirical character; for instance, the form and size of the atomic basis set in a LCAO-type development is scarcely ever determined from first principles.

Spectroscopically accurate solutions of the Schrödinger equations have been computed only for the H_2 molecule in the ground state [5] and several excited states [6]. Chemically accurate results are now available for very simple molecules, like LiH [7]. For most diatomic molecules and hydrides of the first row, there are only moderately accurate calculations [8,9]; it should be noted that extensive computational work is needed to obtain even such accuracy, because electron correlation has explicitly to be taken into account. *Ab initio* calculations performed on molecules of small or medium size by the LCAO SCF method fall into the category of crude calculations. In the case of diatomic molecules [10] and linear polyatomic molecules [11], the energies obtained by the LCAO method are very close to the so-called '*Hartree-Fock limit*', *i.e.* to the best energy given by an independent particle model, but are hardly of moderate accuracy as compared to the experimental values. In the case of a medium-sized molecule, like benzene, the total energy given by the best LCAO SCF calculation available at the present time is equal to -230.46 *a.u.* [12]; the Hartree-Fock limit can be estimated to about -231 *a.u.* and the experimental value to -232 *a.u.* Thus, the best theoretical energy differs from the experimental one by about 1000 kcal·mol^{-1}, and in this respect the 'Quest of the Hartree-Fock limit' is somewhat vain.

Ab initio calculations have two serious shortcomings: they require a high programming effort and are time-consuming, even with fast computers; they have too poor a precision with respect to experiment if limited to the independent-particle scheme. Semi-empirical approaches have been developed for overcoming these difficulties: some simplifications are introduced into the mathematical formalism of quantum-chemical methods, particularly when computing the matrix elements of the Hamiltonian; experimental data are used to calibrate certain quantum-mechanical parameters in order to ensure agreement between theory and experiment. The calculation methods generally used for σ electrons in polyatomic molecules are related to the molecular orbital theory; they are bond orbital or delocalized molecular orbital methods. The former are applied mainly to saturated molecules, because completely localized orbitals seem to offer a good starting assumption for studying additivity rules and their deviation in such compounds. The latter are used in various forms (semi-empirical and non-empirical) for more general electron systems; they are described in Sect. 6.2.

The analysis of observed binding energies [13,14,15] and other properties (molecular refractivities, magnetic susceptibilities, etc...) of saturated hydrocarbons shows that it is possible to reproduce the experimental data to a high degree of accuracy, by adding up a number of the C—C and C—H bonds. A simple additivity scheme giving bond energies to a few kcal·mol^{-1} is unable to distinguish between the various isomers of a molecule, but is significantly improved if different empirical values are attributed to bonds starting from a primary, secondary or tertiary carbon atom [16,17,18], or if interaction parameters between bonded or non-bonded atoms are taken into account [19,20,21,22]. Additivity rules and quasi-constancy of bond lengths, force constants and other properties in saturated molecules are difficult to understand within an *ab initio* framework: they are interpreted by postulating that the σ bonds can be described in terms of electron pairs [23].

If the molecular wave function is approximated by a separated-pair function of the form

$$\Psi(1,2,\ldots,2m) = \mathscr{A}\left[\Psi_A(1,2) \ldots \Psi_K(2k-1,2k) \ldots \Psi_M(2m-1,2m)\right]$$

$$(6.1)$$

where Ψ_K is either an electron pair localized on a bond or a lone pair (see Sect. 4.1), the constants of additivity rules can be identified with certain properties of the separated pairs, provided the bond-interaction terms are small. However, these assumptions do not lead in a straightforward way to additivity rules nor does the existence of additivity rules justify the form (6.1) postulated for the total wave function. Additivity rules could be constructed anyway if the interaction terms were large, but connected with each other by some relations [24]. In an actual *ab initio* calculation on methane [25] it was found that, in spite of a good localization of electrons, the total interpair correlation energy was about 60 kcal·mol^{-1}. A relationship between additivity and localizability can be simply accepted as a reasonable assumption, since these additivity rules break down in systems with a strong delocalization.

If the Coulomb interaction between electrons of different pairs is ignored, each localized bond and lone pair contribute independently to the total energy, which implies a perfect additivity of bond energies. In the independent-particle model, the localized bond function can be visualized as a two-center molecular orbital occupied by two electrons. Nevertheless, it is possible to reproduce deviations from additivity rules within this scheme, for instance, by taking into account overlap (for a review, see *e.g.* [2]).

Chemical evidence shows that in a chain of localized C—C bonds an atom with a different electronegativity induces *intramolecular charge transfers* which are transmitted along the chain (Schlüssel-Atoms of classical organic chemistry [26]). The propagation of inductive effects along a set of two-center molecular orbitals can be introduced in a simple MO-LCAO scheme as follows [27]. Each individual bond is described as the antisymmetrized product of two spin orbitals corresponding to one bond orbital, which is approximated as a linear combination of two atomic orbitals, one for each of the two atoms X and Y participating in the σ bond. The coefficients of the linear combination are the elements of the eigenvector corresponding to the lower eigenvalue of a 2×2 Hamiltonian matrix, whose elements are three empirical parameters: the two diagonal elements α_X and α_Y and the off-diagonal elements $\beta_{XY} = \beta_{YX}$. They play the same role as the α's and β's of the Hückel method. The interaction of one bond with the others is introduced by assuming that each of the α's depends on all the neighboring atoms according to the equation

$$\alpha_X = \overset{0}{\alpha}_X + \sum_Z \gamma_{XZ}\, \alpha_Z \qquad (6.2)$$

where Z denotes any neighbor of X, including Y. For instance, the carbon atom of methane has

$$\alpha_C = \alpha + 0.137\,\beta$$

whereas the carbons of the methyl and methylene groups of ethylamine have

$$\alpha_1 = \alpha + 0.133\,\beta$$

$$\alpha_2 = \alpha + 0.154\,\beta$$

respectively, as compared to

$$\overset{0}{\alpha}_C = \alpha + 0.07\,\beta$$

for a carbon in the absence of neighbors. (The fixed parameters α and β are the zero point and the unit of energy, respectively; β is close to -5 eV [28], which gives values of the order of 0.1 eV for the energy changes brought about in a bond by the environment).

The structure of Eq. (6.2) clearly shows that the effect of the next-nearest neighbors on the α value of a given atom is proportional to the corresponding α through a quantity of the second order in the γ's, as should be expected of inductive effects according to the chemical definition.

For halogenated paraffins [27] and aminoacids [28] this simple method gives a charge distribution which is useful for interpreting molecular properties related to atomic electron densities: dipole moments, quadrupole coupling constants, chemical shifts, etc. . . It can be also applied to the σ framework of heteroaromatic molecules in connection with π-electron calculations [29,30] and yields in an inexpensive way a total $(\sigma + \pi)$ charge distribution analogous to that of more complicated $\sigma - \pi$ calculations (see Sect. 6.4).

There is one property of saturated compounds, which cannot be predicted by additivity rules, namely *ionization potentials*. As has been shown by Lennard-Jones and Hall [31,32], this fact is not inconsistent with the description of saturated molecules in terms of localized bonds and can easily find a place in such a picture.

Consider, for instance, the ionization of methane. Starting from a localized description of the neutral molecule, one can assume that the lowest state of the positive ion is obtained by removing one electron to one of the C−H bond functions, and construct in this way four possible functions of the form

$$\Psi_+ = \mathscr{A}\left[\Psi_A(1,2)\, \Psi_B(3,4)\, \Psi_C(5,6)\, \Psi_D(7)\right] \tag{6.3}$$

$\Psi_D(7)$ denoting the bond function occupied by one electron. The four functions Ψ_+ correspond to the same physical situation and form a set of four degenerate functions with the same energy. Consequently, a better description of the ionized states of methane should be obtained by taking a linear combination of the four functions Ψ_+ [b]. The same formalism applies if one bond is excited instead of being ionized: the four possible excited functions Ψ_{exc} give rise to two distinct states, a triply degenerate one (the lowest excited state of CH_4) and a singly degenerate one. The important result is that the excitation is no longer localized on a particular bond, but distributed on all the bonds. From a theoretical point of view, the preceding treatment is nothing but a configuration interaction limited to four equivalent functions, and it should be made even if the various bonds are not geometrically equivalent, for instance, in long paraffins, because their energies are still very close. In conclusion, the picture of a neutral molecule in terms of localized bonds for the ground state is consistent with extensive delocalization in the upper states.

A *semi-empirical calculation method* for ionization potentials has been developed, using the fact that a Slater determinant is defined only up to a unitary transformation (see Sect. 4.4): the canonical molecular orbitals φ_i, eigenvectors of the Hartree-Fock operator F for a closed-shell system, can be replaced by equivalent orbitals ω_i, almost completely localized,

---

[b] In the frame of the separated-group function formalism, it is found that the four possible functions Ψ_+ are unable to combine together and the ground state of the ion CH_4^+ keeps its fourfold degeneracy [33].

and *vice-versa*. Whereas the operator F is diagonal with respect to the φ functions:

$$\int \varphi_j^* F \, \varphi_i \, d\tau = e_i \, \delta_{ji} \tag{6.4}$$

(see Eq. 2.16), it has diagonal and off-diagonal elements with respect to the ω functions:

$$\alpha_i = \int \omega_i^* \, F \, \omega_i \, d\tau \qquad \beta_{ji} = \int \omega_j^* \, F \, \omega_i \, d\tau \tag{6.5}$$

(note that F is invariant in a unitary transformation). Now, if one identifies, the equivalent orbitals with two-center localized molecular orbitals, *i.e.* one neglects the 'tails' in the equivalent orbitals, one can assume in first and good approximation that the α's and β's depend only on the nature of the bonds directly involved. Furthermore, the β's corresponding to two bonds without a common atom can be considered as exceedingly small. From this, a semi-empirical method may be developed in which the α's and β's are regarded as adjustable parameters. Then, by diagonalizing the α and β, it is possible to estimate the values e_i of the delocalized molecular orbitals φ_i, which are related to ionization potentials *via* the Koopmans theorem (see Sect. 5.2). For instance, the study of paraffins involves two diagonal elements $\alpha_{CC'}$ and α_{CH} and three off-diagonal elements $\beta_{CC',CC''}$ $\beta_{CC',CH}$ and $\beta_{CH,CH'}$; these parameters are fitted (with some simplifications) on the ionization potentials of the first members of the paraffin series and carried over into the whole set of compounds [32]. The equivalent orbital method has been extended to saturated and unsaturated compounds with functional groups (ketones, acids, etc.) and gives surprisingly good agreement with experimental ionization potentials [34]. This agreement is not an indication that the method of equivalent groups is theoretically very sound, but that it is a good interpolation scheme for experimental data. In fact it is not a true theoretical treatment of ionization phenomena, because of the large number of approximations it involves.

Finally, the molecular orbitals φ_i themselves can be expressed as linear combinations of the bond orbitals ω_j (LCBO method):

$$\varphi_i = \sum_j c_{ij} \, \omega_j, \tag{6.6}$$

the coefficients c_{ij} being the components of the eigenvector associated to the eigenvalue e_i. The square of the coefficients in the molecular orbital occupied by a single electron in the positive ion also gives the distribution of the positive charge due to the missing electron in the neutral molecule.

The fraction of positive charge c_{ij}^2 on the jth bond is expected to be correlated with the fragmentation of the ion in mass spectroscopy experiments [35,36]. Other LCBO methods following the same lines have occasionally been used (see e.g. [21]).

6.2. Joint Treatment of σ and π Electrons in Unsaturated Molecules

Two problems arise right from the beginning when one wishes to compute molecular wave functions:

the number of electrons to be explicitly included in the total wave function:

the choice of an expansion basis set.

In large molecules, it is tempting to limit the calculations to valence-shell electrons, i.e. the $1s$ electron for hydrogens, the $2s$ and $2p$ electrons for atoms of the first row and so on, because only those electrons are involved in the usual theories of the chemical bond. Indeed, the semi-empirical extensions of the molecular orbital method suggested for the σ electrons of unsaturated compounds are essentially valence-shell treatments. Some difficulties in the molecular-orbital method originate from that restriction, for instance, the orthogonality problem with respect to the inner shells (see Sect. 3.1) and the definition of core integrals taking into account the attraction of nuclei and repulsion of inner electrons. Moreover, the analysis of important physical properties, for which the inner electrons are largely responsible, has to be done in an indirect way; this is the case for the relationship between K-shell ionization potentials and intramolecular charge transfers [37], or the variation of nuclear spin-spin constants, like $J_{13_{C-H}}$, which depend among other things on a high power of the effective nuclear change of inner orbitals [38]. In principle, all the difficulties should disappear if all the electrons are taken explicitly into account.

In an *ab initio* calculation, the expansion basis set is completely defined by its mathematical form, size and location in space, the orbitals being either centered on the nuclei or floating in more general theories (see e.g. [39]). It does not matter whether pure or hybridized orbitals are used, because nothing is altered by a linear transformation among the members of a basis set, neither the value of the various observables, nor the difficulty of the calculations. On the other hand, in semi-empirical calculations, the atomic orbital basis is chosen with a view to the approximations to be introduced later, and linear transformations of the hybridization type may be useful for transferring some experimental properties into the theory. Obviously, semi-empirical methods can hardly be invariant against such transformations. In the following, the methods available for treating σ and π electrons simultaneously have been classified according to

i) the type of basis (*i.e.* use of pure or hybridized orbitals)

ii) the contents of the one-electron effective Hamiltonian (*i.e.* either the electronic repulsion is completely neglected, or the matrix giving the molecular orbitals contains interaction terms between the electrons explicitly considered in the calculation)

iii) the approximations concerning the integrals themselves (*i.e.* overlap and differential overlap are neglected or taken into account by some means or other).

All-valence electron methods recently suggested for organic molecules are derived from semi-empirical approaches developed earlier in another context, namely the Wolfsberg-Helmholz treatment of coordination compounds [40], the Sandorfy treatment of paraffins [41] and the Parr-Pariser-Pople treatment of π electrons [42,43].

The present discussion has been limited to the most widely used methods, but many variants have been suggested; Table 10 gives a brief survey of them. A more complete report can be found elsewhere [44] c).

For some purposes, a basis set consisting of hybridized atomic orbitals is particularly suitable in LCAO$-$MO calculations. By taking hybrids directed along the chemical bonds instead of pure atomic orbitals defined in terms of arbitrary axes, one simultaneously retains the essential features of the bond orbital picture and the standard delocalized method. This method has been developed in a parametric form similar to the standard Hückel method including or not including overlap integrals [41].

In the case of *hydrocarbons*, the calculation still comprises five parameters, namely two Coulomb integrals α_C and α_H, two bonds, resonance integrals $\beta_{CC'}$ and β_{CH} and one resonance integral for two orbitals centered on the same carbon β_{CC}. As usual, the interaction terms between non-bonded atoms are neglected. The parameters α and β are the matrix elements of a non-specified effective Hamiltonian with respect to the sp^3 or sp^2 hybrid orbitals of carbon and the $1s$ orbitals of hydrogens. For the σ bonds of conjugated hydrocarbons [45], the following set of values has been used

$$\alpha_C = \alpha \qquad \alpha_H = \alpha - 0.2\beta_0$$

$$\beta_{CC'} = \beta \qquad \beta_{CH} = 0.94\beta \qquad \beta_{CC} = 0.38\beta$$

---

c) Digital computer programs are needed for most of the σ-π electron methods; many of them can be obtained through the Quantum Chemistry Program Exchange (Chemistry Department, Indiana University).

Table 10. *Classification of current σ—π electron methods*

Basis set	No explicit electron interaction	Simulated electron interaction	Approximate electron interaction	Full electron interaction
Localized orbitals	Equivalent molecular orbital method[31,32] Group orbital method[34] Two-center molecular orbital method[27]		Second-order perturbation method[124]	
Hybridized orbitals	Hybridized orbital Hückel theory[41,45] Independent electron molecular orbital theory[48]		Zero-differential overlap approximation[46] Orthogonalized SCF method[71]	
Pure atomic orbitals	Extended Hückel theory[40,50] Iterative extended Hückel theories[60 61 62]	Quasi SCF diagonal element method[65] Kinetic-energy Hückel theory[58]	Zero-differential overlap approximation[51,69]	Standard SCF method[73,74,75, 76,77,78] SCF group function method[33] Random-phase approximation[120] Second-order perturbation method[127]

This purely parametric method can easily be converted into a Parr-Pariser scheme, as done for paraffins [46], by selecting appropriate values for the penetration integrals $(U_R^0; pp)$ and electron repulsion integrals $(pp; qq)$ between hybrid orbitals.

It is possible to derive instructive information concerning the general features of the electronic structure of σ bond systems by making drastic assumptions on the value of the parameters α and β. First, one obtains a bond orbital scheme if all resonance integrals between hybrids not involved in a chemical bond are set equal to zero; consequently, the delocalization effects can be treated by the standard perturbation theory of the molecular-orbital method; second, if all the Coulomb integrals are assumed to have the same value and the resonance integrals β_{CC} for orbitals centered on the same atom are neglected (which means that the same value is assigned to the diagonal elements of the 2s and 2p orbitals of carbon), certain results well known in the Hückel theory of alternant hydrocarbons (pairing of the occupied and virtual molecular orbitals, uniformity of charge distribution, sign alternation of mutual polarizabilities etc. [47]) will apply also to all-valence electron treatments [48]. In addition to the value assigned to the parameter of the 1s orbital of hydrogens, the energy difference between 2s and 2p orbitals of carbon plays an important role in the polarity found in the C−H bonds and induces delocalization effects which combine with those produced by long-range resonance integrals. The resulting bond orders were expected to be related to long-range nuclear spin-spin coupling constants, but the general properties of independent particle models suggest rather that the latter are for the most part genuine correlation effects [49].

Nowadays, the success of the methods proposed by Hoffmann [50] and by Pople and Segal [51] among the chemists tends to promote the use of pure atomic orbital bases for all-valence treatments. The first method is a straightforward application of the Wolfsberg-Helmholz treatment of complexes to organic compounds and is called the '*Extended Hückel Theory*' (EHT), because its matrix elements are parametrized in the same way as the Hückel method with overlap for π electrons. The other method, known under the abbreviation '*Complete Neglect of Differential Overlap*' (CNDO), includes electron repulsion terms by extending to σ orbitals the successful approximation of zero-differential overlap postulated for π electrons.

In Extended Hückel Theory, the diagonal matrix elements α_p of the effective Hamiltonian are identified with the corresponding valence-state ionization energies, *i.e.* for carbon and hydrogen atoms:

$$\alpha_{2s}^{C} = -21.4 \text{ eV} \qquad \alpha_{2p}^{C} = -11.4 \text{ eV} \qquad \alpha_{1s}^{H} = -13.6 \text{ eV}$$

and the off-diagonal elements are calculated by one of the following formulas [40,50,52,53]:

$$\beta_{pq} = -K S_{pq} \qquad \text{(overlap proportionality rule, } K = 21 \text{ eV)} \quad (6.7)$$

$$\beta_{pq} = K S_{pq} \frac{\alpha_p + \alpha_q}{2} \qquad \text{(arithmetic mean rule, } K = 1.75) \quad (6.8)$$

$$\beta_{pq} = K S_{pq} \sqrt{(a_p \times \alpha_q)} \qquad \text{(geometric mean rule, } K = 2.00) \quad (6.9)$$

where the K parameters are simple proportionality factors and the overlap integrals are calculated from Slater orbitals with the usual screening constants. The molecular orbitals are obtained by solving a set of Hückel-type equations in which all overlap and resonance integrals are in general retained. Irrespective of their physical meaning, the above formulas include a rather imprecise empirical parameter (plausible values of K for the arithmetic or geometric mean rule range from 1 to 4.5) and have several inconsistencies:

i) They are not invariant with respect to linear combinations of basis orbitals and to shifts in the zero-point of the energy scale, except for very particular transformations [54].

ii) The arithmetic mean rule gives the same resonance integrals β_{pq}, regardless of whether the energies α_p and α_q associated to the orbitals χ_p and χ_q are close to each other or differ from the mean value by an arbitrary quantity. For two orbitals involved in a chemical bond, the geometric mean rule overestimates the covalent bond energy, and in addition has to be taken as an absolute value in order to avoid imaginary resonance integrals β_{pq} due to a possible opposite sign of the integrals α_p and α_q. In this respect the reciprocal mean rule [55]

$$\beta_{pq} = K \, S_{pq} \frac{\alpha_p \, \alpha_q}{(\alpha_p + \alpha_q)/2} \tag{6.10}$$

should work better, since it predicts a bonding power in agreement with the best empirical measure of the covalent bond [56].

iii) All the rules imply that the kinetic T_{pq} contained in each β_{pq} is proportional to the overlap integral S_{pq}, although the numerical values of the kinetic terms are known to vary as the square of the overlap, at least for certain choices of the AO basis. If the right dependence is introduced into the arithmetic mean rule, one obtains the formula

$$\beta_{pq} = S_{pq} \, (2 - |S_{pq}|) \, \frac{\alpha_p + \alpha_q}{2} \tag{6.11}$$

where the absolute value $|S_{pq}|$ is required by the angular factor of the $2\,p$ orbitals [57]. This question is settled once for all in the 'Kinetic-energy-included Extended Hückel Theory' [58,59], where the matrix elements of the kinetic operator T are computed theoretically and only the potential part V of the effective Hamiltonian is evaluated by an arithmetic mean rule:

$$\beta_{pq} = T_{pq} + V_{pq}$$

$$V_{pq} = K \, S_{pq} \frac{V_{pp} + V_{qq}}{2} \tag{6.12}$$

However, better results are obtained if specific values K_{pq} are taken, *i.e.* different proportionality factors for different pairs of atoms and orbitals, and a special formula is used for one-center off-diagonal elements (for details, see [58]). The K factors are not fitted on empirical data but obtained by trial and error after *ab initio* SCF calculations on model compounds (for ethylene, $K_{CC} \simeq 1.0$ for valence-shell orbitals, $K_{HH} = 1.18$, $K_{2s,H} = 1.05$, $K_{2p,H} = 0.98$).

In fact, none of these points is really so important in a semi-empirical method, because such methods are not designed for performing absolute calculations on single molecules, but rather for studying the trends of physical properties in a series of related compounds. Computational experience shows that the general picture of the electronic structure is not significantly altered whatever method may be chosen, as long as the parameters or the coordinate axes of orbitals are varied within reasonable limits.

More serious problems arise with the *energy parameters* α_p. In principle, the valence-state ionization potentials approximating the diagonal elements of the effective Hamiltonian should be selected in accordance with the formal atomic charges; one is necessarily led to an iterative calculation resembling the well-known ω-technique for π electrons, *i.e.* one guesses a charge distribution, corrects the α's for their charge dependence, calculates a new charge distribution from the molecular orbitals found by solving the secular equation and so forth (see *e.g.* [59]). Several extended Hückel schemes of this type including damped iterative procedures [60,61, 62,63] have been proposed, but the problem is not so simple as in π-electron theories, because the α's depend on the formal atomic charges *via* the separated electronic populations of the orbitals located on each atom. It should be remembered that, for instance, the ionization energies of the neutral nitrogen atom are not the same for the valence states $s^2 p_x p_y p_z$ and $s p_x^2 p_y p_z$. An additional difficulty comes from the fact that the orbital populations are not integral numbers (0, 1 or 2) but fractions of an electron, so that one has to define differential ionization energies [64]. From a strictly theoretical point of view, the problem can be settled only if the matrix elements of semi-empirical methods are considered as an approximation for the matrix elements of the SCF one-electron Hamiltonian

$$F = H^{\text{core}} + \sum_i [n_i J_i(1) - {}^1/_2\, n_i K_i(1)] \tag{6.13}$$

where n_i is the occupation number of the molecular orbital φ_i (see Sect. 2.4). This can be done by using in a systematic way the Mulliken-Ruedenberg approximation for the two-electron integrals $(pq;rs)$ contained in the matrix elements of the Coulomb and exchange effective operators J_i and K_i [65]. The following expression is then found for the diagonal elements α_p of the orbital χ_p belonging to the atom P:

$$F_{pp} = W_p + \sum_{r \in P} q_r^p (J_{pr} - {}^1/_2 K_{pr})$$
$$+ \sum_{L \neq P} [-(U_L^0; pp) - \sum_{l \in L} (n_l^L - q_l^L)(J_{pl} - {}^1/_2 K_{pl})]$$
(6.14) [d]

in which q_r^P or q_l^L denote the gross atomic populations of the orbitals χ_r or χ_l respectively centered on the atom P or L, *i.e.* expressions of the following form (see Sect. 6.3):

$$q_r^p = \sum_i \sum_s n_i c_{ir} c_{is} S_{rs} \quad \text{(sum over any s)}, \tag{6.15}$$

and n_l^L is the number of electrons occupying the orbital χ_l in the appropriate valence state configuration of the atom L. Assuming that the population of the orbital χ_l is not too much altered from the atom to the molecule (*i.e.* $n_l^L \simeq q_l^L$) and neglecting penetration integrals $(U_L^0; pp)$, one finds

$$\alpha_p = W_p + \sum_{r \in P} q_r^p (J_{pr} - {}^1/_2 K_{pr}) \tag{6.16}$$

which is the basic formula for a SCF-like extended Hückel theory. The core parameters W_p and one-center two-electron integrals J_{pr} and K_{pr} can be evaluated from the spectroscopic data available for the atom P and its ions; typical values are given in Table 11.

Clearly, the simplified form of α is only valid for almost neutral molecules. In the case of strongly polar molecules and ions, the last term of Eq. (6.14) has to be taken into account, at least through its long-range components J_{pl}^L. It may be remarked that if one introduces the two-center Coulomb integrals in their asymptotic form

$$J_{pl}^L = (pp; ll) \simeq \frac{1}{R_{PL}} \quad \text{(in $a.u.$)} \tag{6.17}$$

one obtains the correction term to be added in a molecular orbital model to account according to Jørgensen for the Madelung energy between the different groups of a molecule [67,68].

---

[d] Actually, Eq. (6.14) is only correct for closed-shell systems, where n_i is equal to 2 for occupied orbitals and 0 for virtual orbitals. It is extended to open-shell systems with $n_i = 1$ for singly occupied orbitals in the Longuet-Higgins and Pople approximation of the Roothaan SCF equations [66].

Table 11. *Atomic parameters for iterative extended-Hückel theories*[1])

Slater-Condon parameters (eV)	H	C	N	O
W_s	—13.59	—51.25	—76.23	—100.71
W_p		—41.83	—61.81	—84.10
J_{ss}	12.85	11.73	13.97	15.11
J_{sp}		11.48	13.65	15.14
J_{pp}		11.51	13.71	15.87
$J_{p_x p_y}$		10.22	12.05	13.77
K_{sp}		2.59	3.05	3.66
$K_{p_x p_y}$		0.64	0.83	1.05

[1]) Calculated from valence state ionization potentials and electroaffinities: G. Pilcher and H. A. Skinner, J. Inorg. Nucl. Chem. *24*, 937 (1962).

A further step in the way of improvements is to consider all the parameters, the α's as well as the β's, as approximate expressions of the SCF effective Hamiltonian. This was done using various zero-differential-overlap approximations [51,69)e]. The diagonal elements F_{pp} of the CNDO method are given by an expression completely equivalent to Eq. (6.14) and the off-diagonal elements are of the form

$$F_{pq} = K_{pq}^{core} S_{pq} - \sum_i \frac{n_i}{2} c_{ip} c_{ip} (pp;qq) \qquad (6.18)$$

In order to preserve the invariance of charge distributions under rotation of the local coordinate axes of each atom, the integrals $K_{pq}^{core} S_{pq}$ and $(pp;qq)$ are assumed to be independent of the azimuthal quantum number of atomic orbitals, *i.e.* the same value is used for any 2 s and 2 p orbitals. Finally, it should be noted that in the case of σ electrons the zero-differential-overlap approximation cannot be justified as completely as for π electrons by arguing about orthogonalized Löwdin orbitals, because the expression of the $S^{-1/2}$ matrix cannot be limited to first-order terms [70,71,72].

Requiring more and more rigor in the computational method inevitably results in the carrying out of *ab initio* calculations by the molecular-

---

e) A short description of the various forms of Zero-Differential-Overlap approximations (ZDO approximation): CNDO 1, CNDO 2, NDDO, INDO, PNDO, ENZDO, MINDO... requires at least a family tree. See also G. Klopman and B. O'Leary: Fortschr. Chem. Forschg. *15*, 445 (1970), „All-Valence Electrons S.C.F. Calculations".

orbital method. At the present time, such calculations are currently made on organic molecules of medium size (*i.e.* containing one ring), using LCAO expansions in atomic orbitals of Gaussian form [73,74,75,76] and also on simple polyatomic molecules with Slater-type atomic orbitals [77,78]. Programs running on large digital computers are necessary for rigorous computation of all the integrals from their mathematical definition and performance of the SCF iterative cycles. A comparison between molecular orbital energies obtained by various semi-empirical and *ab initio* methods for ethylene, formaldehyde and benzene is given in Tables 12, 13, 14.

It may be asked: *do these methods have any practical use* and, if so, what are the merits of the more sophisticated treatments with respect to the simpler ones, for instance, the primitive Extended Hückel Theory? First, quantum-chemical calculations are concerned with the electronic structure and related physical properties. It has been verified that the general picture of the charge distribution is the same in the various calculation methods, especially in the case of heterocyclic compounds [79]. The fine details of the electronic structure have been successfully correlated with certain physical properties: dipole moments, quadrupole coupling constants, chemical shifts, nuclear spin-spin coupling constants, hyperfine coupling constants in free radicals... (see *e.g.* [80]). It is not easy to define the physical meaning of such correlations in the case of highly parametrized methods, but it is gratifying to see that a more satisfactory agreement with experiment may be found by iterative methods (see *e.g.* [81,82]) for nuclear magnetic resonance phenomena). *Ab initio* calculations are needed for an analysis in terms of quantum-mechanical observables: in principle, the mean value of a one-electron operator (*i.e.* the position vector $\vec{r}$ of an electron for dipole moments) is more easily calculated than energy by SCF independent particle models, because the first-order correlation correction vanishes by virtue of the Brillouin theorem [83].

Most *problems of chemical interest* (relative stability of conformers, rotation barriers, equilibrium constants, etc.) involve variations of the total energy rather than one-electron operator mean values. Approximate methods are by definition unable to give any value for the total energy, because they do not explicitly take into account the electron repulsion terms, except for CNDO-type methods. Of course, the molecular orbital energies e_i can be correlated to ionization potentials (see Sect. 5.2), but a sum of ionization energies cannot be identified with the total energy of a SCF scheme. The orbital energies e_i of a closed-shell system are given by

$$e_i = I_i + G_i \qquad (6.19)$$

Table 12. *MO-energies for ethylene*

	Total energy (a.u.)	K shells of C		σ MO's					π,π^* MO's		Ref.
		$1a_g$	$1b_{3u}$	$2a_g$	$2b_{3u}$	$1b_{2u}$	$3a_g$	$1b_{1g}$	$1b_{1u}$	$1b_{2g}$	
Nonempirical LCAO.SCF calculations	−78.01	−305.83	−305.78	−28.29	−21.66	−17.82	−15.81	−14.00	−10.17	+3.91	a)
	−78.00	−305.68	−305.64	−28.09	−21.73	−17.58	−15.91	−13.78	−10.00	+4.04	b)
	−77.95	−306.09	−306.05	−28.31	−21.61	−17.62	−15.48	−13.88	−9.96	+4.83	c)
	−77.85	−307.15	−307.12	−27.16	−20.85	−17.00	−14.93	−13.45	−9.78		d)
	−77.83	−307.13	−307.11	−27.60	−21.29	−17.52	−15.28	−13.77	−10.09	+6.60	e)
	−76.77	−306.26	−306.25	−28.52	−21.36	−15.75	−13.03	−12.05	−10.53	+4.20	f)
Semiempirical calculations				−26.98	−20.60	−16.21	−14.45	−13.78	−13.22		g)
				−24.09	−19.70	−15.65	−13.75	−13.98	−12.24		h)
				−26.01	−19.48	−17.15	−13.01	−10.98	−10.93	+5.28	i)
					−19.18	−15.25	−12.94	−12.76	−10.86		j)
Experiment 1)	−78.62				−19.13	−19.63	−14.39	−12.50	−10.48		c) k)

1) Ionization energies from photoelectron spectroscopy (k)

a) Schulman, J. M., Moskowitz, J. W., Hollister, C.: J. Chem Phys. *46*, 2759 (1967).
b) Buenker, R. J., Peyerimhoff, S. D., Whitten, J. L.: J. Chem. Phys. *46*, 2029 (1967).
c) Moskowitz, J. W.: J. Chem. Phys. *43*, 60 (1965).
d) Rajagopal, P.: Z. Naturforsch. *22a*, 295 (1967).
e) Palke, W. E., Lipscomb, W. N.: J. Am. Chem. Soc. *88*, 2384 (1966).
f) Dierksen, G., Preuss, H.: Intern. J. Quant. Chem. *1*, 365 (1967).
g) Hoffmann, R.: J. Chem. Phys. *39*, 1397 (1963).
h) Yonezawa, T., Yamaguchi, K., Kato, K.: Bull. Chem. Soc. Japan *40*, 536 (1967).
i) Clark, P. A., Ragle, J. L.: J. Chem. Phys. *46*, 4235 (1967).
j) Dewar, M. J. S., Klopman, G.: J. Am. Chem. Soc. *89*, 3089 (1967).
k) Al Joboury, M. I., Turner, D. W.: J. Chem. Soc. 4434 (1964).

Table 13. *MO-energies for formaldehyde*

Total energy (a.u.)	Orbital energies in eV of MO's								π, π* MO's		Ref.
	K shells of O and C										
	$1\,a_1$	$2\,a_1$	$3\,a_1$	$4\,a_1$	$1\,b_2$	$5\,a_1$	$2\,b_2$	$1\,b_1$	$2\,b_1$		
Non empirical LCAO SCF calculations											
−113.89	−559.83	−308.65	−38.20	−23.53	−18.76	−17.70	−11.98	−14.53			a)
−113.83	−560.27	−309.18	−38.91	−23.58	−19.10	−17.52	−12.03	−14.57	+2.93		b)
−113.67	−560.72	−309.04	−38.92	−23.43	−18.76	−17.19	−11.62	−14.25	+3.99		b)
−113.45	−560.24	−309.01	−37.26	−22.77	−18.35	−15.53	−10.48	−12.78	+6.71		c), d), e)
−113.43	−561.19	−310.27	−38.03	−22.62	−18.39	−16.14	−10.76	−13.53	+6.12		e), d)
Semi empirical calculations											
			−36.65	−28.47	−21.20	−17.24	−13.74	−14.92	+0.42		f)
			−30.91	−20.92	−16.73	−14.44	−10.95	−15.28			g)
			−34.06	−22.15	−17.84	−14.85	−13.86	−14.28			h)
Experiment −114.55											
						−13.1	−10.8	−11.8			i)
				−21.	−16.9	−16.0	−10.86	−14.4			j)

a) Neuman, D. B., Moskowitz, J. W.: J. Chem. Phys. *50*, 2216 (1969).
b) Winker, N. W., Dunning, T. H., Letcher, J. H.: J. Chem. Phys. *49*, 1871 (1968).
c) Foster, J. M., Boys, S. F.: Rev. Mod. Phys. *32*, 303 (1960).
d) Newton, M. D., Palke, W. E.: J. Chem. Phys. *45*, 2329 (1966).
e) Aung, S., Pitzer, R. M., Chan. S. I.: J. Chem. Phys. *45*, 3457 (1966).
f) Jungen, M., Labhart, H., Wagniere, G.: Theoret. Chim. Acta *4*, 305 (1966).
g) Carroll, D. G., Vanquickenborne, L. G., McGlynn, S. P.: J. Chem. Phys. *45*, 2777 (1966).
h) Yonezawa, T., Yamaguchi, M., Kato, H.: Bull. Chem. Soc. Japan *40*, 536 (1967).
i) Sugden, T. M., Price, W. C.: Trans. Faraday Soc. *44*, 116 (1948).
j) Brundle, C. R., Turner, D. W.: Chem. Commun. 314 (1967).

Table 14. *MO-energies for benzene*

	Total energy (in a.u.)	K shells of C[1]	σ MO's $2a_{1g}$	$2e_{1u}$	$2e_{2g}$	$3a_{1g}$	$2b_{1u}$	$1b_{2u}$	$3e_{1u}$	$1a_{2g}$	π,π* MO's $1a_{2u}$	$1e_{1g}$	$1e_{2u}$	$1b_{2g}$	Ref.
Non-empirical LCAO-SCF calculations	−230.46	−307.3	−31.78	−28.22	−23.02	−20.08	−18.01	−17.19	−16.92	−14.26	−14.56	−10.15	+ 3.46	+ 9.88	a)
	−229.70	−310.8	−33.27	−29.48	−23.97	−20.83	−18.67	−18.88	−17.78	−15.19	−16.03	−11.62	+ 2.26	+ 8.68	b)
	−227.27	−306.3	−31.62	−28.05	−21.85	−16.62	−16.45	−13.20	−13.99	−10.80	−13.06	− 8.49	+ 4.79	+10.91	c)
	−220.07	−293.2	−26.93	−24.24	−19.42	−15.53	−15.31	−12.20	−13.04	−10.18	−12.30	− 7.83	+ 5.23	+11.16	d)
Semi-empirical calculations			−29.57	−25.78	−19.93	−16.58	−16.60	−14.30	−14.64	−12.84	−14.51	−12.80	+ 3.94		e)
			−31.18	−14.81	−20.10	−19.64	−14.26	−13.81	−12.99	− 9.79	−15.15	− 9.40		+ 7.39	f)
					−18.98	−16.08	−15.67	−13.45	−12.86	−11.54	−12.72	−10.15			g)
Experiment								−16.84			−11.48	− 9.24			h)
						−18.75	(−14.44)	−16.73	−13.67	(−12.19)	−11.49	− 9.25			i)
					−20.26	−18.22	(−15.54)	−16.84	(−14.87)	−13.88	−11.51	− 9.25			j)
					−19.2	−16.9	−15.4	−10.85	−13.8	−10.35	−11.50	− 9.24			k)
					−19.2	−16.9	−15.4	−14.7	−13.8	−11.4	−12.1	− 9.3			l)

1) Mean energy of the six deepest molecular orbitals
2) Ionization energies from Rydberg spectra (h), photoelectron spectroscopy (i,j,k), mass spectrometry (l)

a) Schulman, J. M., Moskowitz, J. W.: J. Chem. Phys. 47, 3491 (1967).
b) Praud, L., Millie, Ph., Berthier, G.: Theoret. Chim. Acta, 11, 169 (1968).
c) Dierksen, G., Preuss, H.: Intern. J. Quant. Chem. 1, 357 (1967).
d) Schulman, J. M., Moskowitz, J. W.: J. Chem. Phys. 43, 3287 (1965).
e) Hoffmann, R.: J. Chem. Phys. 39, 1397 (1963).
f) Clark, P. A., Ragle, J. L.: J. Chem. Phys. 46, 4235 (1967).
g) Dewar, M. J. S., Klopman, G.: J. Am. Chem. Soc. 89, 3089 (1967).
h) El Sayed, M. F., Kasha, M., Tanaka, Y.: J. Chem. Phys. 34, 334 (1961).
i) Al Joboury, M. I., Turner, D. W.: J. Chem. Soc. 1964, p. 4434.
j) Clark, I. D., Frost, D. C.: J. Am. Chem. Soc. 89, 244 (1967).
k) Lindholm, E., Jonsson, B. Ö.: Chem. Phys. 1, 501 (1967).
l) Momigny, J., Lorquet, J. C.: Chem. Phys. Letters 1, 505 (1968).

where I_i and G_i are the matrix elements of the core Hamiltonian and the effective electronic potential contained in the Fock operator (Eq. 2.16). Starting with this expression, several forms can be written for the total energy; for instance

$$E_{SCF} = \sum_i (I_i + e_i) + N \qquad (6.20)$$

$$E_{SCF} = \sum_i (2 e_i - G_i) + N \qquad (6.21)$$

where the sum is to be taken over the doubly occupied molecular orbitals. Therefore, by simply adding the energies of electrons e_i one ignores the fixed-nuclei repulsion N and counts the electronic interaction energy twice [84]. However, it has been suggested [58] that the binding energy A, *i.e.* the difference between the total energy of a molecule E_{SCF}^m and that of the component atoms E_{SCF}^a, could be predicted by means of molecular and atomic orbital energies e_i^m and e_i^a alone, because the quantity

$$\Delta = \sum_i (I_i^m + I_i^a) + N \qquad (6.22)$$

is usually a small part of A (in ethylene, $A_{SCF} = 0.734$ *a.u.*, $\Delta = 0.102$ *a.u.*).

Then, one is justified in putting

$$A = \sum_i (e_i^m - e_i^a) \qquad (6.23)$$

$$E_{SCF} = \sum_i e_i^m + C \qquad (6.24)$$

the quantity C, of purely atomic origin, being constant for a series of isomers. In the case of free radicals, the expression (6.23) should be supplemented by an extra term equal to $1/4 \, J_{rr}$ (the self-interaction of the unpaired electron in the molecular orbital φ_r); this term, derived from the form of the effective Hamiltonian in the SCF theory of Longuet-Higgins and Pople for open-shell sytems, does not seem to be important in discussions of the relative energies of free radicals [81,85]. The binding energies predicted by the preceding formulas are comparatively correct; however, it should be recalled that the binding energies calculated from SCF non-empirical calculations, using the same orbital basis for the molecule and its components, are much smaller than the experimental values (about 40% for aromatic molecules [86]). One-electron theories are considered to be fully reliable only for the *study of angular deformations* and break down completely in the case of very polar molecules [87], because they only take into account the forces associated with the overlap of orbitals, but not the long-range forces coming from the Coulomb interaction of electrons [88]. At the same time, a theoretical justification is found for the empirical correlation diagrams of Mulliken-Walsh which relate the molecular shape to the angular energy variation of the orbitals available for the electrons [89]. The calculation of equilibrium bond lengths and energy derivatives (force constants) is not so successful, even in approximate SCF methods including electron repulsion terms. In addition to the attractive term usually ascribed to overlap, the length of a bond is determined by other factors [88], and its evaluation requires a well-balanced mixture of all the contributions from electrons and nuclei. The approximate methods of CNDO-type have been parametrized to give acceptable values for heats of formation [90] or electronic transition energies [91] for molecules in their actual geometries; since the latter may not correspond to the minimum of the approximate theoretical energy, there is not much hope of obtaining good results for non-equilibrium quantities [90]. On the other hand, recent *ab initio* calculations suggest that the full SCF method is able to reproduce the geometry of poly-atomic molecules in good detail, for instance, the preferential conformation of two rotating methyl groups [92] and the bond length and force constant of the C—H bond in paraffins [93], or the inversion of the N—H bond in heterocyclic compounds [94]. In any case, the real reason why electron correlation seems to play no role in phenomena of that sort should be investigated.

6.3. Analysis of Charge Distributions and the Meaning of Formal Atomic Charges.

Density contour maps, like those of Fig. 1 (Sect. 2.3) for the nitrogen molecule or Fig. 2 (Sect. 4.2) for ethylene give a complete picture of the electronic distribution in a molecule. However, it is more convenient, especially for comparative studies, to describe the electronic structure by a set of single numbers rather than by maps, even if this involves the loss of much information. This is why indices summarizing the form of the electron distribution in the neighborhood of an atom or a bond have been defined by quantum chemists. Following Mulliken, the assignment of a set of such indices to a molecule may be called its *population analysis*.

There are two principal sorts of population analysis: that of Coulson and Longuet-Higgins, expressed in terms of charges (often called 'charge densities') and bond orders [95], and that of Mulliken in terms of atomic and overlap populations [96]. Both are strictly defined within the frame of the LCAO—MO method.

In the primitive definition of charges and bond orders it was assumed that the atomic basis orbitals are orthonormal. Then, the charge associated with the rth atomic orbital χ_r and the bond order associated with the pair of atomic orbitals χ_r and χ_s are

$$q_r = \sum_i n_i c_{ir}^2 \tag{6.25}$$

$$p_{rs} = \sum_i n_i c_{ir} c_{is} \tag{6.26}$$

where n_i is the occupation number of the molecular orbital φ_i in the state under consideration and the c_{ir}'s are the expansion coefficients of φ_i. Since several atomic orbitals may be centered on an atom, the charge density of an atom P and the bond order between two atoms P and Q are obtained by summing the contributions coming from the various orbitals belonging to them:

$$q_P = \sum_{r \in P} q_r^P \tag{6.27}$$

$$p_{PQ} = \sum_{r \in P} \sum_{s \in Q} P_{rs}^{PQ} \tag{6.28}$$

In independent-particle models, the charge and bond orders are the representation of the first-order density matrix $\gamma(1,1')$ in the basis of the χ functions (see Sect. 2.3). Clearly, the sum of the diagonal elements q_r or q_P is equal to the number of electrons n:

$$\sum_r q_r = \sum_P q_P = n \tag{6.29}$$

and the charge density can be interpreted as the probability of finding an electron close to the atom P.

For extending the preceding definitions to the case of non-orthogonal basis set, two procedures have been devised, which can be reconciled on the basis of a more physical definition of charges based on dipole moments (*vide infra*):

i) The overlap integrals are the components of a metric tensor in an m-dimensional space, where the contravariant coefficients associated to the molecular orbitals φ_i are defined by

$$d_{ir} = \sum_s S_{rs} c_{is} \tag{6.30}$$

the corresponding covariant coefficients being the coefficients c_{ir} of the primitive basis functions χ_r [97]. Then, the charge and bond orders are given by

$$q_r = \sum_i n_i c_{ir} d_{ir} \tag{6.31}$$

$$p_{rs} = \tfrac{1}{2} \sum_i n_i (c_{ir} d_{is} + c_{is} d_{ir}) \tag{6.32}$$

ii) The set of basis functions χ is replaced by the equivalent set of Löwdin orthogonalized orbitals λ:

$$\lambda = S^{\frac{1}{2}} \chi \qquad (6.33)$$

and the coefficients of the molecular orbitals with respect to the new basis functions λ_r:

$$b_{ir} = \sum_s S_{rs}^{-\frac{1}{2}} c_{ir} \qquad (6.34)$$

are put in Eqs. (6. 25) and (6.26), giving charge and bond orders without overlap [58,99].

The population analysis of a LCAO—MO wave function requires three kinds of indices: the atomic populations

$$\varrho_P = \sum_{r \in P} \sum_i n_i c_{ir}^2 \qquad (6.35)$$

the overlap populations

$$\varrho_{PQ} = \sum_{r \in P} \sum_{s \in Q} 2 n_i c_{ir} c_{is} S_{rs} \qquad (6.36)$$

and the gross atomic populations

$$q_P = \varrho_P + \sum_{Q \neq P} \frac{1}{2} \varrho_{PQ} \qquad (6.37)$$

The gross atomic populations are identical with the charge densities including overlap given by Eq. (6.31), but different from the charge densities calculated from orthogonalized atomic orbitals. The expression (6.37) shows that the charge density of an atom P includes contributions coming from non-bonded atoms; at the moment, it is just a formal, but convenient way of distributing the electrons between atoms (see e.g. [96]). By subtracting the charge q_P from the number of electrons n_P contributed by the atom P, one obtains the net electric charge of P (or formal atomic charge) with its conventional sign:

$$\delta_P = n_P - q_P \qquad (6.38)$$

The distribution of net charges in molecules is often visualized in the form of 'charge diagrams' (Fig. 6 and 7). In planar molecules, the charges q_P can be separated in a σ component q_{P_σ} and a π component q_{P_π}; hence

σ and π net charges can be defined if one is able to say how many σ and π electrons the atom P has contributed to the molecule:

$$\delta_{P_\sigma} = n_{P_\sigma} - q_{P_\sigma}$$

$$\delta_{P_\pi} = n_{P_\pi} - q_{P_\pi}$$

(6.39)

This is generally possible for neutral molecules in their ground states: for an unsaturated carbon, one has $n_{C_\pi} = 1$; for doubly-bonded nitrogen as in pyridine $n_{N_\sigma} = 6$, $n_{N_\pi} = 1$, and for a simply-bonded one as in pyrrole or aniline $n_{N_\sigma} = 5$, $n_{N_\pi} = 2$. In ionized or excited states, there may be some ambiguity concerning the origin of electrons, especially in the case of excited states where the total number of σ and π electrons is not the same as in the ground state (for example, the $n - \pi^*$ excited states of carbonyl compounds).

The values of charge densities and the net charges calculated from them must be accepted with discretion. The atomic orbitals chosen for expanding the molecular orbitals have a considerable effect upon their magnitude. Not only do charge densities depend on the orbital exponents of the atomic orbitals, but they are not invariant with respect to a linear transformation of the orbital basis set, so that their physical meaning may be disputed[f]. Recent calculations made for methane [33,93] and ethane [92,100] clearly show what difficulties arise in their interpretation. If one chooses a Slater minimal basis set using for the exponents of hydrogen $1s$ orbitals the value corresponding to the free atom $\zeta_H = 1,0$, or that of the hydrogen molecule $\zeta_H = 1.2$, the charge on the hydrogens of methane does not vary very much: $\delta_H = +0.131$ or $+0.113$. However, if the exponents of all the orbitals, those of carbon as well as those of hydrogen, are determined by minimizing the total energy of methane itself, the charge transfer from hydrogens to carbon is almost annihilated: $\delta_H = +0.019$. However, the charge of hydrogens is also reduced by simply orthogonalizing the basis set and calculating charges corresponding to the new orbitals: with $\zeta_H = 1.0$, it is found from Eq. (6.25) $\delta_H = +0.085$ instead of $\delta_H = +0.131$ [33].

In Table 15, the net charges obtained by different methods for the hydrogen atoms of various *hydrocarbons* are compared. All the calculations, except the semi-empirical ones involving a special parametrization, give a positive charge on the hydrogens and a negative charge on carbon, whether the hydrogens are linked to a simply-bonded atom (paraffins) or a doubly or triply-bonded atom (ethylene or acetylene).

---

[f] Population indices are invariant with respect to a unitary transformation among doubly occupied molecular orbitals [96] (for instance, with respect to a localization process).

Table 15. *Net charges of hydrogen in various hydrocarbons*

	Non-empirical SCF methods						Semi-empirical SCF methods		
	$Z_H = 1.2$ a)	$Z_{opt.}$ for CH_4 b)	Gaussians c)	$Z_H = 1$[1] 1	2	3	EHT e)	PNNDO f)	Bond MO g)
Methane	+0.113	+0.019		+0.131	+0.085	+0.079	+0.133	−0.077	(0)
Ethane Stg.	+0.124	+0.004					+0.119	−0.064	(0)
Ethane ecl.		+0.003							
Ethylene	+0.140						+0.113	−0.033	+0.057
Acetylene	+0.188						+0.157	+0.059	+0.165
Benzene			+0.208				+0.101		
Pyridine o			+0.222				+0.10		
Pyridine m			+0.217						
Pyridine p			+0.220						

[1] SCF calculations with Slater orbitals (1), orthogonalized Slater (2) and orthogonalized group functions (3) The same basis integrals are used in the three calculations.

a) Palke, W. E., Lipscomb, W. N.: J. Am. Chem. Soc. *88*, 2384 (1966).
b) Pitzer, R. M.: J. Chem. Phys. *46*, 4871 (1967); *47*, 965 (1967).
c) Praud, L., Millie, P., Berthier, G.: Theoret. Chim. Acta *11*, 169 (1968); Clementi, E.: J. Chem. Phys. *46*, 4731 (1967).
d) Klessinger, M., McWeeny, R.: J. Chem. Phys. *42*, 3343 (1965).
e) Hoffmann, R.: J. Chem. Phys. *39*, 1397 (1963).
f) Dewar, M. J. S., Klopman, G.: J. Am. Chem. Soc. *89*, 3089 (1967).
g) Pople, J. A., Santry, D. P.: Mol. Phys. *9*, 301 (1965).

Actually, such a charge distribution is by no means related to the usual polarity of the C—H bonds in physical organic chemistry for instance, the $C^+—H^-$ polarity in methane as opposed to the $C^-—H^+$ polarity in acetylene. The meaning of theoretical charge distributions has recently been clarified by calculating bond moments of equivalent molecular orbitals almost localized on the C—H bonds [101,102]. The C—H bond moments turn out to be of the same order of magnitude (1.8 D for acetylene, 1.9 D for ethylene, 2.0 D for ethane) with the negative end of the electric dipole on the hydrogen atom. This result can be understood by considering a localized bond function constructed from an sp, sp^2 or sp^3 hybrid orbital of carbon and the $1s$ orbital of hydrogen. The center of the negative charge distribution corresponding to a carbon hybrid directed towards the hydrogen atom, does not coincide with the carbon nucleus, but lies *almost in the middle of the C—H bond*. Even if the coefficients of the carbon and hydrogen orbitals are equal, the bond functions C—H will have a dipole moment in the sense $C^+—H^-$. The analysis of molecular wave functions in terms of localized orbitals shows that the moments resulting from the preceding mechanism always prevail for the C—H bonds, in spite of a total charge distribution in the opposite sense. Using the terminology of dipole moment theory, it can be said that 'homopolar dipole moments' of 'hybridization dipole moments' are responsible for the larger part of the C—H bond moment. However, if one wants to calculate the total dipole moment, it may happen that the various hybridization moments cancel more or less and can be ignored in first approximation.

Predicting experimental quantities by means of purely theoretical indices, like charges and bond orders, instead of calculating the expectation values of approximate operators could be criticized on the grounds that population analysis opens the door to a 'plague of non-observables' [103]. Nevertheless, this procedure is to some extent justified by several reasons, in addition to its convenience. First, it is found that many observables other than dipole moments can be expressed in terms of charge densities, bond orders and related quantities, if the martix elements of the corresponding operator with respect to the basis functions are approximated in terms of overlap integrals by a Mulliken-type formula. Such is the case with quadrupole coupling constants [104,86], spin-orbit coupling constants [105] and nuclear spin-spin coupling constants [106]. As regards dipole moments, it amounts mainly to neglecting the contributions coming from hybridization moments, provided the orbitals centered on a given atom are chosen to be orthogonal (see below).

A comparison between the values deduced from charge densities and those obtained with the vector position operator, using *ab initio* SCF wave functions of unsaturated heterocycles [107,3], shows that the *dipole moments* calculated from total net charges ($\delta_\sigma + \delta_\pi$) are underestimated: in pyrrole, point charges give 1.22 D, whereas an exact calculation gives 2.10 D, and experiment gives 1.74 D); in pyridine, the corresponding values are 1.89, 3.11, and 2.20 D. Likewise, *quadrupole coupling constants* deduced from $2p$ charge densities are underestimated: in pyrrole, 3.94 MHz from point charges, 5.26 MHz from the exact calculation, 2.06 MHz from experiment; in pyridine, the corresponding values are 5.30, 6.31 and 4.58 MHz respectively [108,109]. In these calculations, the discrepancy between point charge and exact computations might be due in part to the fact that the basis functions centered on the same atom are orthogonal only in the case of orbitals with different symmetries. However, it should be possible to eliminate the corresponding intra-atomic overlap integrals by appropriate linear transformations and obtain better results, for instance, by adding point charge and hybridization dipole moments.

The definition of the point charges to be used for molecular diagrams should be based on the expression of a well-specified physical observable, rather than on an arbitrary albeit intuitively satisfactory partition of MO-LCAO wave function, as is the case with the above definitions. An analysis of this question, with reference to the *electric dipole moment*, has been recently presented [110,111]: instead of calculating the dipole moment from the point charges previously defined, the expression for the net charges is deduced from the quantum-mechanical expression of that observable. The main point is that the electric dipole moment of a molecule can be divided in a unique way into three contributions: hybridization or atomic contribution, overlap contribution, and charge-transfer contribution (which provides a definition of net atomic charges), each being uniquely defined within the MO-LCAO scheme.

The MO-LCAO expression of the dipole moment of a molecule in the chemical convention and in atomic units is

$$\vec{\mu} = \sum_P \sum_Q \sum_{r \in P} \sum_{s \in Q} p_{rs}^{PQ} \int \chi_r^* \vec{r} \chi_s \, dr \tag{6.40}$$

where p_{rs}^{PQ} is defined by an expression similar to Eq. (6.26), namely

$$p_{rs}^{PQ} = \sum_i n_i \, c_{ir} \, c_{is}$$

where n_i is the occupation number of the molecular orbital φ_i, c_{ir} and c_{is} the coefficients of the orbitals χ_r and χ_s belonging to the atoms P and Q, respectively, in the LCAO expansion of φ_i. The general position vector of an electron and the position vector of atom P (all measured with respect to the same arbitrary origin in *a.u.*) are denoted by $\vec{r}$ and $\vec{r}_P$, and n_P is written for the number of electrons with which atom P participates in the formation of the molecule.

One can now carry out the following substitutions

$$\vec{r} = \vec{r}_P + \vec{\xi} \qquad\qquad \text{in those terms where } P = Q;$$

(6.41)

$$\vec{r} = \tfrac{1}{2}\left(\vec{r}_P + \vec{r}_Q\right) + \vec{\eta} \qquad \text{in those terms where } P \neq Q.$$

Evidently, $\vec{\xi}$ is the radius vector taken from the position of atom P, $\vec{\eta}$ is the radius vector from the center of the line PQ. Taking into account the orthogonality of the two orbitals $\chi_r, \chi_{r'}$ belonging to the same atom P, calling S_{rs}^{PQ} the overlap integral of orbitals centered on different atoms P and Q and letting

$$\vec{\xi}_{rs}^{P} = \int \chi_r^* \vec{\xi} \chi_s \, d\tau \qquad\qquad \vec{\eta}_{rs}^{PQ} = \frac{\int \chi_r^* \vec{\eta} \chi_s \, d\tau}{S_{rs}^{PQ}}$$

(6.42)

the dipole moment (6.40) becomes

$$\vec{\mu} = \sum_P \delta_P \vec{r}_P + \vec{\mu}_{\text{hybrid}} + \vec{\mu}_{\text{overlap}}$$

(6.43)

where δ_P is the net charge of atom P:

$$\delta_P = \sum_{r \in P} \left(p_{rr}^{PP} + \sum_{Q \neq P} \sum_{s \in Q} p_{rs}^{PQ} S_{rs}^{PQ} \right) - n_P$$

(6.44)

Since the hybridization moment defined by

$$\vec{\mu}_{\text{hybrid}} = \sum_P \sum_{r,r' \in P} p_{rr'}^{PP} \vec{\xi}_{rr'}^{P}$$

(6.45)

can be set equal to zero if each atom contributes only orbitals of the same symmetry and the overlap moment

$$\vec{\mu}_{\text{overlap}} = \sum_P \sum_{r \in P} \sum_Q \sum_{s \in Q} p_{rs}^{PQ} S_{rs}^{PQ} \vec{\eta}_{rs}^{PQ}$$

(6.46)

vanishes if the centroids of the various pairs of orbitals coincide with the centers of the corresponding P-Q lines (*i.e.* $\vec{\eta}_{rs}^{PQ}$ is zero) and/or if their differential overlap is negligible, it follows that Eq. (6.44) gives the charges we were looking for. It is evident that the definition (6.44) is perfectly consistent with formula (6.38). Comments on the transformation of this formula upon orthogonalization of the atomic orbital basis are given in Ref. [111].

6.4. Interaction of σ and π Charge Distributions

The interaction of σ and π electrons can be described in various forms, according to the method used to construct the molecular wave function. The most easily visualizable one is the electrostatic interaction of σ and π charge distributions in an independent particle model. That sort of interaction is included in all-electron SCF calculations through the effective Fock Hamiltonian (2.23). If we confine ourselves to the linear or plane molecules, it is possible to divide the total electron density into σ and π parts, coming respectively from σ and π occupied orbitals and giving rise to q_{P_σ} and q_{P_π} populations of Eq. (6.39). Starting from the σ and π components of the electron density, it is also possible to define σ and π dipole moments under the same conditions as net charges δ_{P_σ} and δ_{P_π} (see Sect. 6.3). For symmetry reasons, both components of the total dipole moment are directed along the internuclear axis or lie in the molecular plane, but there is no reason why they should have the same sense. Actually, recent calculations suggest than σ and π charge transfers in heteropolar molecules may be opposite, as a result of the σ–π electrostatic interaction [112,113].

Consider, for instance, hydrogen cyanide HCN: the σ charge is preferentially attracted by the nitrogen atom, which is more electronegative than carbon; consequently, the π electrons are subjected to the effect of differently charged atoms and try to balance the σ charge distribution. The actual charge distribution in HCN can be considered as a result of a 'two-way charge transfer'[112], nitrogen being the most negative center for the σ system, as is predicted from electronegativity considerations, and carbon the most attractive atom for the π system, in contradiction to the assumption of the standard π theory. Of course, the overall charge distribution yields a dipole moment with its negative end towards the nitrogen atom (Fig. 6).

Such opposite polarities of σ and π systems resulting in a small total polarity are well known in transition metal complexes and are expressed by the electroneutrality principle of the ligand-field theory [114]. Roughly speaking, similar things are to be expected in organic molecules having a doubly-bonded heteroatom, as nitrogen in pyridine [113]. However, ab initio calculations on aromatic aza-compounds [107] rather suggest that the π polarity of carbon and nitrogen atoms is not reversed, as in the HCN molecule, but simply compensated by the effect of the σ charge distribution (Fig. 7).

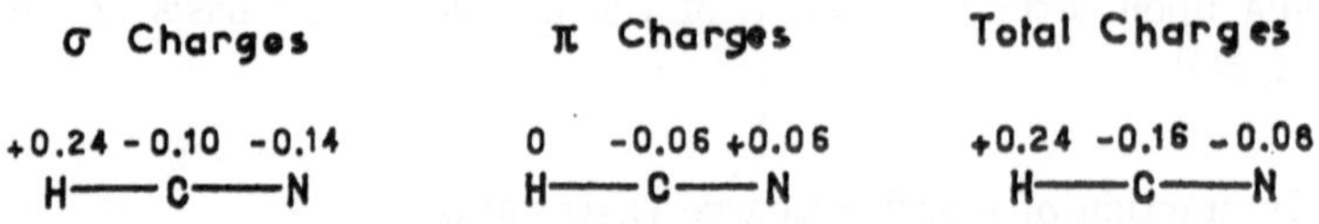

Fig. 6. Charge diagrams of hydrogen cyanide

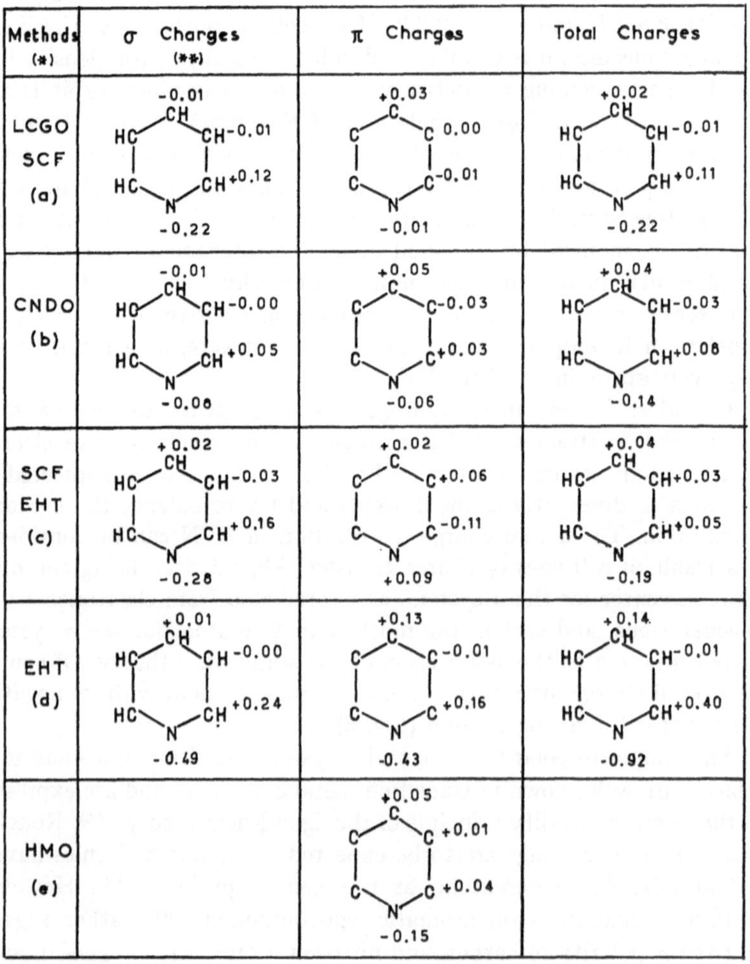

Fig. 7. Charge diagrams of pyridine from various calculations

Clearly, the π electron cloud of pyridine looks much more like that of benzene than is usually postulated in the Hückel theory. As a matter of fact, the dipole moment of 3.11 D (exp 2.20 D), calculated from the SCF wave function of ref. [107] using the dipole length operator, includes a very small π component (0.325 D) oriented in the same direction as a large σ component (2.785 D) [115) g].

The $\sigma-\pi$ interaction taken into account by independent particle models of the SCF type arises also in pure hydrocarbons. A striking example is given by the methyl radical and its ions: the charge transfer from hydrogen to carbon in $C-H$ bonds decreases from the positive ion, CH_3^+ to the neutral radical CH_3 and the negative ion CH_3^-, as the number of π electrons on the carbon atom passes from 0 to 1 and 2; in other words, increasing the π-electron density on carbon produces a lowering of its actual electronegativity with respect to the σ electrons [116].

Charge diagrams obtained from different calculation methods are generally in good agreement with each other, except for the magnitude of the charge transfer within certain bonds, such as $C-H$ or $N-H$ bonds. However, the net charges of atoms in those bonds have a rather limited sense, and in the diagrams of pyridine given in Fig. 7 only the sum of the atomic net charges in each $C-H$ bonds is indicated. As a general rule in heterocycles, the main features of the electronic structure obtained by complete all-electron treatments and rough $\sigma-\pi$ calculations are very similar, and the direction and magnitude of total dipole moments predicted from them are almost identical [117]. Furthermore, the total charge of nitrogen in pyridine-like molecules agrees in an astonishing way with the charges computed by the simple Hückel method; the reason why a calculation limited to π electrons simulates to some extent the results of a $\sigma-\pi$ calculation is probably that the total energy is not very much affected by a small change in partition of total charges into σ and π components [118].

In most molecules, it is possible to describe the $\sigma-\pi$ interaction by simple electrostatic considerations and to explain in this way physical properties depending on them like dipole moments. Electron correlation seems to play no role, except for molecules with a very small polarity, like carbon monoxide [119]. The matter is more complicated for excitation phenomena, because it is necessary to take into account possible changes in the charge distribution, even if the electronic structure

---

g) The dipole moment of molecules with simply-bonded heteroatoms, such as pyrrole nitrogen or furan oxygen, may include large σ and π components, resulting from the fact that these heteroatoms bear a net negative σ charge, because of their larger electronegativity, and a net positive π charge, because of the delocalization of their $2p\pi$ lone pair [86].

of the initial and final states could be understood within the frame of an independent-particle model. In fact, the σ charge distribution in most unsaturated molecules is different from that in isolated atoms, and pure π-electron calculations based on unperturbed valence-state potentials of the GMS type (see Sect. 5.3) can hardly include σ effects involving both the ground state and the excited state. The use of a fixed GMS potential could be completely justified only for alternant aromatic molecules, where there is no extensive σ charge transfer in the ground and π–π^* excited states. This gives an explanation of the fact that semi-empirical π-electron theories give a very satisfactory description for the spectra of aromatic hydrocarbons. The same sort of treatment is far from being so succes ful in the case of $\pi \rightarrow \pi^*$ transitions in highly polar molecules, such as carbon dioxide [120] or the pyridinium ion [121]. However, it can be much improved simply by modifying the potential felt by the π electrons according to the ground-state σ charge distribution obtained in a separate approximate calculation [120,121]. The σ-π charge interaction has a more marked effect on $n \rightarrow \pi^*$ transitions, and the occupied molecular orbitals by the σ electrons have to be explicitly considered for electronic transitions between σ and π levels, for instance, the $n \rightarrow \pi^*$ transitions of oxygen compounds like acrolein and furan [118].

Until quite recently the role of σ–π correlation effects was ignored in the theoretical treatment of electronic transitions. Even now, nearly all *ab initio* calculations of excitation phenomena are based on independent-particle models using a minimal basis set of atomic orbitals, or involve a configuration interaction limited to the π-electron system. In order to go far enough beyond the σ–π separation, two improvements have to be simultaneously considered:

i) a configuration interaction involving both σ^* and π^* virtual molecular orbitals;

ii) a more flexible basis set containing a larger number of atomic orbitals than the standard minimal basis set.

Numerical computations on the lowest singlet-singlet and singlet-triplet transition of ethylene (Table 16) suggest that either type of improvement could produce a better agreement with experiment. Consider the lowest $N \rightarrow V$ transition of ethylene: using the SCF molecular orbitals built for standard Slater atomic orbitals ($1s$, $2s$, $2p_x$, $2p_y$, $2p_z$), the transition energy is found to be equal to 11.98 eV [122,123], that is to say, 4.4 eV above the experimental value. This much too high value is further increased by one eV by a configuration interaction within the π molecular-orbital system (see Sect. 5.3); on the other hand, it is reduced to 10.17 eV and 9.44 eV by performing a σ–π configuration interaction which includes all the single and double excitations with respect to the ground-

Table 16. π—π^* transitions of ethylene from $(\sigma+\pi)$ ab initio calculations

Basis set	SCF	CI limited to π electrons	All-electron CI limited to				Exp. (eV)	Ref.
			Single excitations		Double excitations			
			TDA[1]	Perturb[2]	RPA[1]	Perturb[2]		
Transition $^3B_{2u}\leftarrow{}^1A_g$								
Slater $Z_H = 1.2$	3.45	4.55	3.19	3.40				a)
Slater $Z_H = 1.0$	3.36		3.36	3.30	imaginary		4.6	b) c)
Extended Gauss.	4.14					5.54		d)
Transition $^1B_{2u}\leftarrow{}^1A_g$								
Slater $Z_H = 1.2$	12.10	13.20	10.17 (f = 0.73)	9.75				a)
Slater $Z_H = 1.0$	11.98 (f = 1.03)		8.19 (f = 0.65)		9.44 (f = 0.51)		7.6 (f ≃ 0.3)	b) c)
Extended Gauss.	9.31 (f = 0.95)				7.71 (f = 0.48)	10.78		d)

1) Tamm-Dankoff or Random-Phase approximation (ref. b)
2) Perturbation calculation including the configurations singly excited with respect to the ground-state configuration A_g or the mono- and diexcited configurations with respect to both basis configurations A_g and B_{2u} (Ref. c).

a) Kaldor, U., Shavitt, J.: J. Chem. Phys. 48, 191 (1968).
b) Dunning, T. H., McKoy, V.: J. Chem. Phys. 47, 1735 (1967).
c) Malrieu, J. P., Levy, B., Berthier, G.: (unpublished results).
d) McKoy, V.: Private communication (conference at the Faculté des Sciences d'Orsay 1968).

state configuration [122]. However, if the configurations that are singly and doubly excited with respect to the basic excited configuration are taken into account when calculating the energy of the excited state itself the latter values are brought back to 10.78 eV [124]. Now, with an extended basis set including several $2p\pi$ atomic orbitals per carbon, the primitive SCF result is much lower: 9.31 eV instead of 11.98 eV, and configuration interaction reduces this value to 8.19 eV or 7.71 eV, if singly or doubly excited configurations with respect to the ground configuration are included [125]. Nevertheless, the last value is probably not the end of the story, since the configurations doubly excited with respect to the basic excited configuration have not been included.

An approximate treatment for taking into account $\sigma-\pi$ interaction has been developed in the case of long polyenes [126], and non-empirical calculations have been carried out for the various transitions of formaldehyde along the same lines as for ethylene [127,128,129] [h]. In view of the intricacies of theoretical considerations concerning excited states, it is rather fortunate that calculations limited to the π-electron systems can be forced to agree with experiment by introducing semi-empirical corrections on well-chosen matrix elements.

6.5. Quasi-π and Quasi-σ Orbitals

The distinction between σ and π orbitals is rigorously justified only in planar molecules, where the former are symmetric and the latter are antisymmetric with respect to the molecular plane. This distinction is also meaningful for locally planar systems if the orbitals can be localized in such a way that they are confined to a planar region: a classification with respect to the local symmetry is then possible. An example is a saturated chain with a phenyl group at either end: the two phenyl groups are practically independent of each other, so that there are two independent π systems.

Obviously, the distinction between σ and π orbitals cannot break down completely if there is a slight deviation from planarity, e.g. for a planar molecule in the course of an out-of-plane vibration. Thus, it can be useful to distinguish quasi-σ and quasi-π orbitals. In an LCAO-MO description the quasi-π orbitals are linear combinations of $2p$ atomic

---

[h] The transition energies quoted from refs. [122,127] have been calculated using the techniques of theoretical physics known as the *Tamm-Dankoff Approximation* (TDA) and *Random-Phase Approximation* (RPA). These approximations can be considered equivalent to a CI treatment limited to singly excited configurations (TDA) or doubly excited configurations (RPA) with respect to the ground configuration. (For a critical study of the RDA approximation in the case of imaginary energy transitions, see [130,131]).

orbitals, the axes of which are not strictly parallel. Well-known examples of molecules with quasi-π systems are biphenyl and other biaryls.

In (hypothetical) planar biphenyl the molecular plane is clearly the symmetry plane with respect to which the σ orbitals are symmetric and the π orbitals are antisymmetric. When the two rings form a small dihedral angle, one can still define a *quasi-π system*, provided that the orbitals of the twisted system can be considered as slight modifications of those of the planar biphenyl system. The quasi-π orbitals in question have properties close to those of π orbitals proper, in particular as regards delocalization.

For a dihedral angle of 90⁰ we have to deal with two independent quasi-π systems perpendicular to each other. *A priori* there is no reason why a quasi-π system should not extend into the region of the other ring and be 'conjugated' with the MO's of the other ring that have the same symmetry (with respect to the symmetry group D_{2h}), in particular, with the appropriate linear combinations of the σ orbitals of the nearest atoms on the other ring. Certain observed features of perpendicular biphenyl seem to suggest that some *ring-ring conjugation* does in fact exist: the well-known red shift of sterically hindered biphenyls with respect to benzene is the most important piece of evidence in this connection. However, a careful analysis shows that the red shift is not conclusive evidence of conjugation, because it may explained by different mechanism [132,133,134] involving exciton-type or other long-range interactions. A detailed account of this problem and its bearing on the definition of quasi-π electrons can be obtained by consulting Refs. [135,136, 137,138]. Other examples of molecules with possible 'conjugation' between perpendicular planar subunits are unsaturated spiro-compounds [139,140].

This type of conjugation, the quantitative importance of which is difficult to assess, is closely related to the problem of *hyperconjugation*. A discussion of hyperconjugation is beyond the scope of the present review (see *e.g.* [141,142]. Its importance for the σ—π separation problem lies in the fact that, whenever hyperconjugation plays a major role, the usual rule according to which π systems are associated with double bonds and planar molecules breaks down. Other examples of systems where delocalization extends beyond a conjugated doubly-bonded system are given by cyclopropyl derivatives, where quasi-π orbitals of the cyclopropyl group have often been introduced, at least in qualitative arguments (see *e.g.* [143]).

The notion of quasi-π orbitals is probably quite useful for the description of *reaction intermediates* or of molecular systems in the course of a chemical reaction. In fact, certain reactions can be described by the transformation of a π orbital into a σ orbital or *vice versa*, the other

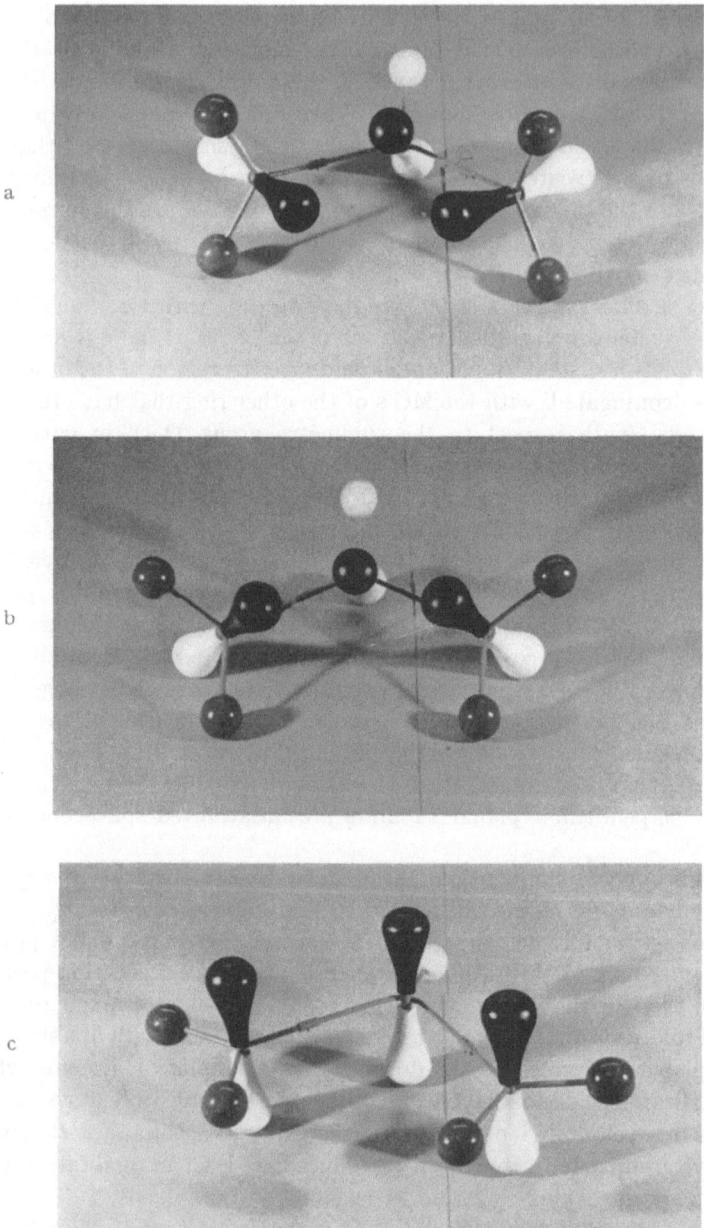

Fig. 8a—c. Rearrangement of Orbitals in the Reaction Cyclopropyl ⇄ Allyl Cations

orbitals being relatively unaffected [144,145,146]. A simple example is the isomerization of the cyclopropyl cation to the allyl cation [147]. The transformation of the π orbital of the former into a σ orbital of the latter is shown in a crude pictorial way in Fig. 8. The molecule with a quasi-π system, roughly on the middle of the reaction path, is normally not a stable species but just one point of the potential energy hypersurface. However, there is evidence that in some cases, *e.g.* in bicyclic systems [148], a stable intermediate with a quasi-π system like that of Fig. 8 is formed as a result of particular steric conditions.

6.6. References

[1] Heitler, W.: Intern. J. Quant. Chem. *1*, 12 (1967).
[2] Klopman, G.: Tetrahedron *19*, Suppl. 2, 111 (1963).
[3] Mely, B., Pullman, A.: Theoret. Chim. Acta *13*, 278 (1969).
[4] Clementi, E., André, J. M., André, M. Cl., Klint, D., Hahn, D.: Acta Phys. Acad. Sci. Hung. *27*, 493 (1969).
[5] Kołos, W., Wolniewicz, L.: J. Chem. Phys. *41*, 3674 (1964), *49*, 404 (1968).
[6] — — J. Chem. Phys. *43*, 2429 (1965), *48*, 3672 (1968).
[7] Boys, S. F., Handy, N. C.: Proc. Roy. Soc. (London) *A 311*, 309 (1969).
[8] Bender, C. F., Davidson, E. K.: J. Phys. Chem. *70*, 2675 (1966).
[9] Ahlrichs, R., Kutzelnigg, W.: Theoret. Chim. Acta. *10*, 377 (1968).
[10] Cade, D. E., Huo, W. M.: J. Chem. Phys. *47*, 614, 649 (1967).
[11] McLean, A. D., Yoshimine, M.: IBM J. Res. Dev. *12*, 206 (1968).
[12] Schulman, J. H., Moskowitz, J. W.: J. Chem. Phys. *47*, 3491 (1967).
[13] Fajans, K.: Ber. Deutsch. Chem. Ges. *53*, 1922 (1920).
[14] Rossini, F.: Chem. Rev. *27*, 1 (1940).
[15] Pitzer, K. S.: J. Chem. Phys. *8*, 711 (1940); J. Am. Chem. Soc. *70*, 2140 (1948).
[16] Proser, E. J., Johnson, W. H., Rossini, F. D.: J. Res. Natl. Bur. Stand. *37*, 51 (1946).
[17] Sonders, M., Matthews, C. S., Hurd, C.O.: Ind. Eng. Chem. *41*, 1048 (1949).
[18] Franklin, J. L.: Ind. Engl. Chem. *41*, 1070 (1949).
[19] Platt, J. R.: J. Chem. Phys. *15*, 419, (1947); J. Phys. Chem. *56*, 328 (1952).
[20] Allen, T. L.: J. Chem. Phys. *31*, 1039 (1959).
[21] Pitzer, K. S., Catalano, E.: J. Am. Chem. Soc. *78*, 4844 (1956).
[22] Bartell, L. S.: J. Chem. Phys. *32*, 827 (1960).
[23] Allen, T. L., Shull. H.: J. Chem. Phys. *35*, 1644 (1961).
[24] Dewar, M. J. S., Pettit, R.: J. Chem. Soc. 1644 (1954).
[25] Ahlrichs, R., Kutzelnigg, W.: Phys. Letters *1*, 651 (1968).
[26] Eistert, B.: Chemismus und Konstitution, Stuttgart: Enke 1948.
[27] Del Re, G.: J. Chem. Soc. 4031 (1958).
[28] Yonezawa, T., Del Re, G., Pullman, B.: Bull. Chem. Soc. Japan *37*, 985 (1964); Biochem. Biophys. Acta *75*, 153 (1963).
[29] Berthod, H., Pullman, A.: J. Chim. Phys. *62*, 942 (1965).
[30] — Giessner-Prettre, C., Pullman, A.: Theoret. Chim. Acta, *8*, 212 (1967).
[31] Lennard-Jones, J. E., Hall, G. G.: Discussions Faraday Soc. *10*, 18 (1951).
[32] Hall, G. G.: Proc. Roy. Soc. (London) *A 205*, 541 (1951).
[33] Klessinger, M., McWeeny, R.: J. Chem. Phys. *42*, 3343 (1965).

34) Franklin, J. L.: J. Chem. Phys. 22, 1304 (1954).
35) Fueki, K., Hirota, K.: Nippon Kagahu Zasshi 81, 212 (1960).
36) Hall, G. G.: Cong. Coll. Univ. Liège 30, 5 (1964).
37) Siegbahn, K., Nordling C. et al.: In: ESCA, Nova Acta Reg. Soc. Sci. Ups. Ser IV, 20, 76 (1967).
38) Barbier, C., Berthier, G.: Theoret. Chim. Acta. 14, 71 (1969).
39) Hurley, A. C.: Intern. Quant. Chem. 1S, 677 (1967).
40) Wolfsberg, M., Helmholz, L.: J. Chem. Phys. 20, 837 (1952).
41) Sandorfy, C.: Can. J. Chem. 33, 1337 (1955).
42) Pariser, R., Parr, R. G.: J. Chem. Phys. 21, 466, 767 (1953).
43) Pople, J. A.: Trans. Faraday Soc. 49, 1375 (1953).
44) Jug, K.: Theoret. Chim. Acta. 14, 91 (1969).
45) Fukui, K., Kato, H., Yonezawa, T., Morokuma, K., Imamura, A., Nagata, C.: Bull. Chem. Soc. Japan 35, 38 (1962).
46) Katagiri, S., Sandorfy, C.: Theoret. Chim. Acta. 4, 203 (1966).
47) Coulson, C. A., Longuet-Higgins, H. C.: Proc. Roy. Soc. (London) A 192, 16 (1947).
48) Pople, J. A., Santry, D. P.: Mol. Phys. 7, 269 (1964); 9, 301, 311 (1965).
49) Barbier, C., Gagnaire, D., Berthier, G., Lévy, B.: J. Magn. Res. (to be published).
50) Hoffmann, R.: J. Chem. Phys. 39, 1397 (1963).
51) Pople, J. A., Segal, G. A.: J. Chem. Phys. 43, S 129, S 136 (1965).
52) Ballhausen, C. J., Gray, H. B.: Inorg. Chem. 1, 111 (1962).
53) Lohr L.L., Lipscomb, W. N.: J. Chem. Phys. 38, 1607 (1963).
54) Berthier, G., Del Re., G., Veillard, A.: Nuovo Cimento 44, 315 (1966).
55) Yeranos, W. A.: J. Chem. Phys. 44, 2207 (1966).
56) Allen, T. L.: J. Chem. Phys. 27, 810 (1957).
57) Cusachs, L. C.: J. Chem. Phys. 43, S 157 (1965).
58) Newton, M. D., Boer, F. B., Lipscomb, W. M.: J. Am. Chem. Soc. 88, 2353, 2361, 2367.
59) Ehrenson, S.: J. Am. Chem. Soc. 91, 3693, 3706 (1969).
60) Carroll, D. G., Armstrong, A. T., McGlynn, S. P.: J. Chem. Phys. 44, 1865 (1966).
61) Rein, R., Fukuda, N., Win, W. H., Clarke, G. A., Harris, F. E.: J. Chem. Phys. 45, 4743 (1966).
62) Ažman, A., Bohte, Z., Ocvirk, A.: Theoret. Chim. Acta, 6, 189 (1966).
63) Duke, B. J.: Theoret. Chim. Acta. 9, 260 (1968).
64) Jørgensen, C. K., Horner, V. M., Hatfield, W. E., Tyree Jr., S. Y.: Intern. J. Quant. Chem. 1, 191 (1967).
65) Berthier, G., Millié, Ph., Veillard, A.: J. Chim. Phys. 62, 8 (1965).
66) Longuet-Huggins, H. C., Pople, J. A.: Proc. Phys. Soc. (London) A 68, 591 (1955).
67) Berthier, G., Lemaire, H., Rassat, A., Veillard, A.: Theoret. Chim. Acta. 3, 213 (1965).
68) de Brouckère, G.: Bull. Soc. Chim. Belg. 76, 407 (1967).
69) Schuster, P.: Monatsh. Chem. 100, 1015, 1033.
70) Klopman, G.: J. Chem. Phys. 43, S 151 (1965).
71) Dahl, J. P.: Acta. Chem. Scand. 21, 1244 (1967).
72) Cook, D. B., McWeeny, R.: Chem. Phys. Letters, 1, 588 (1968).
73) Preuss, H.: Mol. Phys. 8, 157 (1964).
74) a) Whitten, J. L., Allen, L. C.: J. Chem. Phys. 43, 5170 (1965).
    b) Whitten, J. L.: J. Chem. Phys. 44, 359 (1966).
75) Barnett, M. P.: Rev. Mod. Phys. 35, 571 (1963).

76) Clementi, E., Davis, D. R.: J. Comp. Phys. *1*, 223 (1966).
77) Palke, W. E., Lipscomb, W. N.: J. Am. Chem. Soc. *88*, 2384 (1966).
78) Yoshimine, M., McLean, A. D.: Intern. J. Quant. Chem. *1 S*, 313 (1967).
79) Pullman, A.: Jerusalem Symp. Quant. Chem. Biochem. *2*, 9 (1969).
80) Davies, D. W.: The Theory of the Electronic and Magnetic Properties of Molecules, London: Wiley 1967.
81) Ellinger, Y., Rassat, A., Subra, R., Berthier, G.: Theoret. Chim. Acta, *10*, 289 (1968).
82) Berthier, G., Faucher, H., Gagnaire, D.: Bull. Soc. Chim. 1872 (1968).
83) Møller, C., Plesset, M. S.: Phys. Rev. *46*, 618 (1934).
84) Mulliken, R. S.: J. Chim. Phys. *46*, 497 (1959).
85) Gelus, M., Vay, P. M., Berthier, G.: Theoret. Chim. Acta. *9*, 182 (1967).
86) Berthier, G., Praud, L., Serre, J.: Jerusalem Symp. Quant. Chem. Biochem. *2*, 40 (1969).
87) Allen, L. C., Russell, J. D.: J. Chem. Phys. *46*, 1029 (1967).
88) Ruedenberg, K.: Rev. Mod. Phys. *34*, 326 (1962).
89) Peyerimhoff, S. D., Buenker, R. J., Allen, L. C.: J. Chem. Phys. *45*, 734 (1966).
90) Baird, N. C., Dewar, M. J. S.: J. Chem. Phys. *50*, 1262, 1275 (1969); J. Am. Chem. Soc. *91*, 352 (1969).
91) Del Bene, J., Jaffé, H. H.: J. Chem. Phys. *48*, 1807, 4050 (1968); *49*, 1221 (1968); *50*, 1126 (1969).
92) Pitzer, R. M., Lipscomb, W. N.: J. Chem. Phys. *39*, 1995 (1963).
93) — J. Chem. Phys. *46*, 4871 (1967).
94) Veillard, A., Lehn, J. M., Munsch, B.: Theoret. Chim. Acta. *9*, 275 (1968).
95) Coulson, C. A., Longuet-Higgins, H. C.: Proc. Roy. Soc. (London) *A 191*, 39 (1947).
96) Mulliken, R. S.: J. Chem. Phys. *23*, 1833, 1841, 2338, 2743 (1955).
97) Chirgwin, B. H., Coulson, C. A.: Proc. Roy. Soc. (London) *A 201*, 196 (1950).
98) Löwdin, P. O.: J. Chem. Phys. *18*, 365 (1950).
99) McWeeny, R.: J. Chem. Phys. *19*, 1614 (1951); *20*, 920 (1951).
100) Pitzer, R. M.: J. Chem. Phys. *47*, 965 (1967).
101) Gey, E., Havemann, U., Zülicke, L.: Theoret. Chim. Acta. *12*, 313 (1968).
102) Pritchard, R. H., Kern, C. W.: J. Am. Chem. Soc. *91*, 1631 (1969).
103) Platt, J. R.: Handbuch Physik *37/2*, 173 (1961).
104) Guibe, L., Lucken, E. A. C.: Mol. Phys. *10*, 273 (1966); *14*, 73, 79 (1968).
105) Horani, M., Leach, S., Rostas, J., Berthier, G.: J. Chimie phys. *63*, 1015 (1966).
106) Barbier, C., Berthier, G.: Int. J. Quant. Chem. *1*, 657, (1967).
107) Clementi, E.: J. Chem. Phys. *46*, 4725, 4731, 4737 (1967).
108) Schempp, E., Bray, P. J.: J. Chem. Phys. *48*, 2381 (1968).
109) Kochanski, E., Lehn, J. M., Levy, B.: Chem. Phys. Letters *4*, 75 (1969).
110) Del Re, G.: Nuovo, Cimento *17*, 644 (1960).
111) — Intern. J. Quant. Chem. *1*, 293 (1967).
112) Clementi, E., Clementi, H.: J. Chem. Phys. *36*, 2824 (1969).
113) Veillard, A., Berthier, G.: Theoret. Chim. Acta *4*, 347 (1966).
114) Pauling, L.: The Nature of the Chemical Bond, Chap. 5, Ithaca: Cornell University Press 1960.
115) Berthier, G.: Chimia *22*, 385 (1968).
116) Millié, Ph, Berthier, G.: Intern. J. Quant. Chem. *2*, 67 (1968).
117) Pullman, A.: Jerusalem Symp. Quant. Chem. Biochem. *2*, 9 (1970).
118) Jungen, M., Labhart, H.: Theoret. Chim. Acta, *9*, 345, 366 (1968).
119) Grimaldi, F., Lecourt, A., Moser, C.: Intern. J. Quant. Chem. *1 S*, 153 (1967).
120) Hinkelmann, H.: Diplomarbeit, Göttingen (1968).

121) Denis, A., Gilbert, M.: Theoret. Chim. Acta, *11*, 31 (1968).
122) Dunning, T. H., McKoy, V.: J. Chem. Phys. *47*, 1735 (1967).
123) Kaldor, U., Shavitt, I.: J. Chem. Phys. *48*, 191 (1968).
124) Levy, B.: (unpublished results).
125) McKoy, V.: (unpublished results).
126) Denis, A., Malrieu, J. P.: Theoret. Chim. Acta. *12*, 66 (1968).
127) Dunning, T. H., McKoy, V.: J. Chem. Phys. *48*, 5263 (1968).
128) — Hunt, W. J., Goddard III, W. A.: Chem. Phys. Letters *4*, 147 (1969).
129) Malrieu, J. P., Levy, B., Berthier, G.: (unpublished results).
130) Čižek, J., Paldus, J.: J. Chem. Phys. *47*, 3976 (1967); *52*, 2919 (1970); *53*, 821 (1970).
131) Terasaka, T., Matsushita, T.: Chem. Phys. Letters *4*, 384 (1969).
132) Longuet-Higgins, H. C., Murrell, J. N.: Proc. Phys. Soc. (London) *68 A*, 610 (1955).
133) Grinter, R.: Mol. Phys. *11*, 7 (1966).
134) Rastelli, A., Momicchioli, F.: Boll. Sci. Fac. Chim. Ind. Bologna *24*, 189 (1966).
135) Suzuki, H.: Bull. Chem. Soc. Japan *32*, 1340, 1350, 1357 (1959).
136) Jaffé, H. H., Chalvet, O.: J. Am. Chem. Soc. *85*, 1561 (1963).
137) Spetzer, E. G., Jaffé, H. H.: Spectrochim. Acta *23 A*, 1923 (1966).
138) Rastelli, A., Pozzoli, S. A., Del Re, G.: Atti Soc. Nat. Mat. Modena *99*, 233 (1968).
139) Simmons, H. E., Fukunaga, T.: J. Am. Chem. Soc. *89*, 5208 (1967).
140) Hoffmann, R., Imamura, A., Zeiss, G. D.: J. Am. Chem. Soc. *89*, 5215 (1967).
141) Mulliken, R. S.: Tetrahedron *5*, 253 (1959).
142) Dewar, M. J. S., Schmeising, H. N.: Tetrahedron *5*, 166 (1959).
143) Bernett, W. A.: J. Chem. Educ. *44*, 17 (1967).
144) Woodward, R. B., Hoffmann, R.: J. Am. Chem. Soc. *87*, 395 (1965).
145) Longuet-Higgins, H. C., Abrahamson, E. W.: J. Am. Chem. Soc. *87*, 2046 (1965).
146) Fukui, K.: Tetrahedron Letters *24*, 2009 (1965).
147) Kutzelnigg, W.: Tetrahedron Letters *49*, 4965 (1967).
148) Schöllkopf, U., Fellenberger, K., Patsch, M., von Rague Schleyer, P., Su, T., Dine, E. W.: Tetrahedron *49*, 3639 (1967).

7. Conclusions

It is difficult to review the question of σ—π separation without discussing more or less all the aspects of theoretical chemistry; hence the present review may appear too extensive to some and too restricted to others. Among the subjects which we have either barely mentioned or completely ignored, and which yet belong to our topic, are, for instance, aromaticity, antiaromaticity, homo-aromaticity, spiroconjugation, the Woodward-Hoffmann rules, etc.; for these concepts are meaningless unless the σ—π separation is accepted and, indeed, extended to non-planar systems. The popularity among pure chemists of, for example, the Woodward-Hoffmann rules, shows how well rooted the belief in π-electron systems is in present-day chemistry. Our task has been to try to place new emphasis on the way in which this notion is defined and on its limitations and shortcomings in the context of the quantum-mechanical treatment of molecules. This task is especially difficult because, on the other hand, quantum chemists are now drifting away from π-electron theories, whether pure or with allowance for changes in the σ core; and many prefer to carry out all-electron or all-valence electron calculations at different degrees of accuracy. We hope that the present work will help to prevent any confusion arising from the existence of such opposing tendencies.

As far as we can see, any attempt to explain the properties of a molecule by considering explicitly only electrons belonging to a particular class should be encouraged, because idealizations and simplifications are well known to be necessary for an understanding of the physical world. In particular, a full understanding of the properties of organic molecules is greatly facilitated by dividing their electrons into classes, $e.g.$ into σ and π electrons of the entire molecules or of parts of them. This usually amounts to distinguishing the 'mobile' or delocalized electrons from the localized ones, the latter being mainly responsible for the properties of individual bonds, and hence not so relevant when effects involving several bonds, like conjugation, are under study. Therefore, the properties characteristic of unsaturated compounds can (and to some extent should) be described in terms of π electrons in the field of a σ core; indeed one can often apply what is called a 'pure π electron theory', by treating the σ core as if it did not depend on the distribution of the π electrons.

Of course, there must be rules and limitations enabling one to decide when and how such a simplified treatment can be used; and there must be a possibility of comparison with more complete treatments, so that the origin of disagreements may be found and specified. When this is done, there remains the danger of explaining disagreements by introducing terms which serve only to give a name to an otherwise undefined set of neglected effects.

This applies in particular to the so-called $\sigma-\pi$ interaction, which is often introduced generically to explain away the fact that pure π-electron theories sometimes fail to explain facts or give serious quantitative disagreement with experiment.

In fact, the concept of $\sigma-\pi$ interaction has been used in several different and ambiguous ways. Therefore, we close the present review by listing some of the points discussed here which are especially important for clarifying the matter.

1. In planar unsaturated molecules (to which the majority of conjugated systems belong) it is always possible and justifiable to distinguish between σ and π electrons. This distinction can be considered as a 'separation' in the strict sense of the word if it is introduced within the frame of the independent-particle model, because it then becomes possible to define one effective Hamiltonian operator for the σ electrons and one for the π electrons, thus splitting the eigenvalue equation into a system of two equations coupled only through the potentials appearing in the effective Hamiltonians. It is possible to justify the $\sigma-\pi$ separation also in a slightly more general context than that of independent-particle model, but the failure of this separation to explain certain facts must normally be attributed to electron correlation.

2. Some phenomena, like the existence of a hyperfine structure of the ESR spectra of free radicals, which is due to coupling with the proton spins, are completely outside the frame of naive π considerations and should be explained in terms of more general theories. However, most physical properties of molecules are not seriously sensitive to correlation effects; they can be understood in terms of the MO theory, and, in the particular case when the given molecule is planar, within the frame of the $\sigma-\pi$ separation.

For instance, the equilibrium geometries of molecules appear to be determined by the combined action of σ and π electrons. True enough, a pure π electron theory cannot give either equilibrium distances or equilibrium angles. Nevertheless, on the basis of simple π electron calculations and appropriate assumptions regarding the potential of the σ core, several conclusions can be drawn, for instance, regarding the alternation of bond lengths in conjugated polyenes. Furthermore, the fact that benzene is planar, while cyclo-octatetraene is not, can be explained in terms of the competition between the σ and π contributions to the binding energy. The π electrons try, so to speak, to force a planar arrangement, the σ electrons create opposing sterical forces.

3. Even in the frame of an independent-particle model, interaction between the σ and π electrons is taken into account; the potential energy terms appearing in the effective Hamiltonian for π electrons do include terms representing the field created by the σ electrons, and *vice versa*. The additional assumption is often made that changes in the distribution of the π electrons affect the σ electrons so slightly that the potential created by the latter on the former is always the same; sometimes, however, such a pure π electron treatment fails, rather because of this assumption than because of the neglect of the electron correlation.

4. The ionization and excitation phenomena in unsaturated compounds can also be understood, at least qualitatively, in the frame of the $\sigma-\pi$ separation. In fact, the most important absorption bands of organic molecules in the visible and near UV spectral region can be interpreted as arising from $\pi \rightarrow \pi^*$ transitions in the one-electron picture, and hence they can be understood even within a pure π-electron theory involving a rigid σ core. However, especially in compounds containing hetero-atoms, there are transitions which must be interpreted as $\pi \rightarrow \sigma^*$ and $\sigma \rightarrow \pi^*$ transitions; in particular, the excitation of the lone pair leads in general to low-intensity $n \rightarrow \pi^*$ transitions, which also lie in the visible or near UV region. Likewise, the Rydberg series of unsaturated molecules which belong to the $\pi \rightarrow \sigma^*$ type starts in the near UV. In these cases, it is obvious that the differences in the σ cores associated with the different states involved in the transitions under study must be taken into account.

5. Strictly speaking, in the interpretation of spectra, the hypothesis of a rigid σ core is satisfactory only in alternant hydrocarbons because of the negligible horizontal charge shift. In compounds containing hetero-atoms, and also in non-alternant hydrocarbons, even the prediction of $\pi \rightarrow \pi^*$ transitions can be unreliable if no allowance is made for the polarization of the σ core. An even stronger limitation than the assumption of a rigid σ core applies when the potential of the σ core is approximated through the GMS potential. Recent *ab initio* calculations support the reasonable opinion that the GMS potential is an acceptable approximation only in alternant hydrocarbons.

6. In connection with the question of the σ core, two points are important. First, the σ core of unsaturated compounds is not of the same type as in saturated compounds, as is clearly indicated by the difference in hybridization usually attributed to the two classes of compounds. Second, the occupied σ orbitals are often associated with orbital energies. lower than those of the π orbitals, and it is true that the highest occupied orbital of an unsaturated hydrocarbon is a π orbital. However, this does

not mean that all the σ orbital energies lie below the π orbital energies, and the remark just made holds only as a simple possibility. In other words, the fact that the orbital energies of the σ and π orbitals are not separated into two bands is not an indication that the $\sigma-\pi$ separation is not valid.

Acknowledgement: This work is to a large extent the result of a collaboration sponsored by NATO (Research Grant 223), whose support is gratefully acknowledged. The Authors also express their gratitude to the laboratories and institutions which provided the atmosphere and the facilities for it, namely the CNR (Gruppo Chimica Teorica, Roma), the 'Laboratoire de Biochimie théorique associé au CNRS' (Paris), and the 'Lehrstuhl für theoretische Chemie' (Göttingen).

The authors also thank many colleagues for discussions, for reading the manuscript and for making papers available prior to publication; in particular they express their gratitude to their collaborators, R. Ahlrichs, M. Jungen, B. Lévy, J. P. Malrieu, A. Meyer, A. Rastelli and V. Staemmler.

They also thank their colleagues who have given permission to reproduce figures, namely K. Ruedenberg (Fig. 4), H. Preuss (Fig. 2), A. C. Wahl (Fig. 1).

Received November 26, 1970

Springer-Verlag
Berlin · Heidelberg · New York

München · London · Paris ·Tokyo · Sydney

Topics in Current Chemistry

Fortschritte der chemischen Forschung

Schriftleitung: F. Boschke

Band 15, Heft 4:

All-Valence Electrons S.C.F. Calculations

4 fig. 90 pages (pp. 445–534). 1970
DM 38,—; US $ 11.00

Contents: G. Klopman and B. O'Leary, Introduction to All-Valence Electron S.C.F. Calculations of Large Organic Molecules: ·Theory and Applications.

The most important all-valence electron methods proposed for S.C.F. calculations of the properties of large organic molecules are discussed. (Approx. 80 references)

W. Demtröder:

Band 17:

Laser Spectroscopy

16 fig. III, 95 pages. 1971. DM 28,—; US $ 8.10

L. Maier, G. Zon and K. Mislow

Band 19:

The Chemistry of Organophosphorus Compounds I

11 fig. IV, 94 pages. 1971. DM 34,—; US $ 9.90

H. J. Bestmann and R. Zimmermann

Band 20:

The Chemistry of Organophosphorus Compounds II

III, 147 pages. (In German). 1971. DM 58,—
US $ 16.80

In kritischen Übersichten werden in dieser Reihe Stand und Entwicklung aktueller chemischer Forschungsgebiete beschrieben. Sie wendet sich an alle Chemiker in Forschung und Industrie, die am Fortschritt ihrer Wissenschaft teilhaben wollen.

In der Regel werden nur Beiträge veröffentlicht, die ausdrücklich angefordert worden sind. Schriftleitung und Herausgeber sind aber für ergänzende Anregungen und Hinweise jederzeit dankbar. Manuskripte können in den „Fortschritten der chemischen Forschung" in Deutsch oder Englisch veröffentlicht werden.

Jeder Band der Reihe ist einzeln käuflich.

This series presents critical reviews of the present position and future trends in modern chemical research. It is addressed to all research and industrial chemists who wish to keep abreast of advances in their subject.

As a rule, contributions are specially commissioned. The editors and publishers will, however, always be pleased to receive suggestions and supplementary information. Papers are accepted for "Topics in Current Chemistry" in either German or English.

Any volume of the series may be purchased separately.